To Nan.

LIVING
Our
BEST LIVES

THE CANNON HALL FARM STORY

LIVING
Our
BEST LIVES

THE CANNON HALL FARM STORY

MIRROR BOOKS

LIVING OUR BEST LIVES
THE CANNON HALL FARM STORY
by Nicole Carmichael

This book first published in 2021 by Mirror Books

Mirror Books is part of Reach plc
One Canada Square,
London E14 5AP,
England

www.mirrorbooks.co.uk

ISBN 978-1-913406-52-3

First hardback edition 2021

Cover image and photography: Barry Marsden

CONTENTS

Part Two

THE AUTHOR
Nicole Carmichael

NICOLE Carmichael is a national journalist, writer and author. Now living in London, she grew up in York and hopes one day to return home to God's Own Country. Her favourite animals at Cannon Hall Farm are pygmy goats Primrose and Millie and Helen the alpaca.

By Peter Wright
(Yorkshire Vet)

 AS a Veterinary Surgeon of 40 years I have learnt to recognise caring and compassionate farmers and good farming practise. Visiting Cannon Hall Farm, and on meeting The Nicholson Family for the very first time, I knew on both accounts I was in the midst of 'Farming Royalty'.

Cannon Hall Farm is run by Roger with his three sons, Richard, Robert and David. Richard, a keen artist and photographer, works behind the scenes with the marketing of this award-winning farm. Robert and David are the faces of the farm itself and are now household names – even at 10 Downing Street – with their social media insight into current happenings on the farm and through their Channel 5 programmes *Springtime on The Farm* and *This Week on The Farm* based at Cannon Hall.

It is said behind every good man there is a good woman and Roger's wife Cynthia has always been the backbone of this charismatic, dyed in the wool Yorkshire farming family.

Roger was born a farmer from a long lineage who had worked the land around Barnsley before him. As the saying goes, it is in his blood.

Back in Roger's youth he would rush home from school to milk the cows by hand, sitting on a small milking stool, and one evening in October 1956 having settled the animals down for the night, he was faced with the devastating sight of one of the farm buildings containing calves and piglets in full blaze. With no time to raise the alarm Roger rushed in to move the panic stricken animals to safety. This is the Roger Nicholson I have got to know so well where, under any circumstances, the love and wellbeing of his stock comes before anything else. Roger and his family's love of animals shines through on every page.

Roger Nicholson a true Yorkshire man is one of the most likeable, genuine, hardworking and caring men you could ever meet. As this book shows life has not been easy, far from it, but he has succeeded with his family by his side through grit, determination and quiet resolve whilst smiling in the face of adversity. Their legacy will live on for future generations of the family and for the benefit of the wider public, where modern lifestyles often take them further away from farming and the natural world that the Nicholsons nurture. And take out and therein lies the simple attraction of Cannon Hall Farm.

With a young family to support and with Roger and Cynthia working every hour God sent, even with Roger's farming expertise and care of his stock they still struggled to make end meets.

Small family farms in the 1980s were becoming increasingly unprofitable. But Cynthia, also a lady to roll up her sleeves and "get on with it", had an idea to open up a tearoom at the farm, baking cakes and scones and many other treats, but despite working very hard and it being very popular they just about broke even. It needed to extend to make it profitable but there was no money to pay for any extension.

Roger felt if they could earn a little more, with an extended tearoom, there may be enough to support them and so the idea was hatched to diversify from a traditional farm to a tourist attraction.

The family were full of drive and enthusiasm. But the idea was met with a shake of the head from the bank adviser, who pointed out that Roger's overdraft had increased with each of the farming ventures he had tried over the years as each one had failed to turn a profit. Undeterred and with typical tenacity, they secured help from The Yorkshire Bank and Roger and the family took on a siege mentality and an evolve or die mindset.

As the book shows friends and family were called in to help to keep costs down. This help so typifies the Yorkshire spirit and by Easter 1989 the farm was able to open to the public, complete with extended tearoom, including Cynthia's original scone recipe now famous as "The Scone That Saved The Farm".

It is a fact that Cannon Hall continues to evolve. No one in the family sits back on their laurels. It isn't in their nature. Samuel Goldwyn (the film producer) once said "The harder I work the luckier I seem to get," but Roger could never have predicted the rise from the humble homestead they shared, to what is now a major tourist attraction with recognition, respect and fame far beyond the boundaries of Yorkshire.

Roger wanted a financially secure family farm where he could look after his beloved animals, every one. It is after all the family's love of these animals that stands out for all to see, as Cannon Hall continues to evolve.

Meet the Nicholsons

The Nicholsons are the popular farming family featured in the Channel 5 series *Springtime on the Farm, This Week on the Farm* and *Friday on the Farm*.

They come from a farming dynasty that can be traced way back to the 1600s and live at the award-winning Cannon Hall Farm, near Barnsley, in South Yorkshire, a thriving working open farm which plays host to more than 320,000 visitors every year.

Heading up the family is Roger. He has lived at Cannon Hall Farm since 1959 and yet, every time a new breed is introduced or there is a new birth, he is every bit as excited as he was the day he first arrived at the farm as an ambitious 16-year-old. He has been the driving force behind the farm's remarkable diversification from a traditional country smallholding to the huge success story it is today.

Roger's wife, Cynthia, is rather camera-shy, so she doesn't often appear on television, but she is very much the backbone of the family. Also from a farming background, she is happiest when surrounded by her children, grandchildren and great granddaughter, Nelly.

Richard is the eldest of Roger and Cynthia's three sons and heads up the marketing department at Cannon Hall Farm. He is a keen photographer and cook, known for his weekly online cooking demonstrations.

Robert is the middle brother and half of a television farming double act with his younger brother David. He has never wanted to do anything else but farming. He is passionate about animal welfare and loves sharing Cannon Hall Farm's evolving story on social media.

And finally, David, the youngest of Roger and Cynthia's boys, is the JCB loving expert lamber and sheep shearer of the family and is always up for a challenge – especially if he can include some dancing and jokes along the way.

The whole of the Nicholson family is devoted to ensuring that all of the animals on the farm live their best lives and that their animal co-stars – including Jon Bon Pony, Zander the Alpaca and Ozzie Horseborn – continue to delight farm visitors, television audiences and online fans around the world.

Introduction

Several impressive vintage black and white photographs grace the walls of the White Bull Restaurant at Cannon Hall Farm, near Barnsley in South Yorkshire. They each feature a handsome white Shorthorn bull called Sam, whose pedigree name was Fockerby Ringleader.

Beloved by his owner, farmer Charlie Nicholson, Sam the white bull wasn't only best in show at many local farm shows around Yorkshire during the 1950s but was also famous for his taste for stout.

The story goes that whenever Sam was awarded another rosette, Charlie would lead the prize bull to the beer tent, where they would each enjoy a celebratory pint to toast their victory. Sam would obligingly open his mouth and Charlie would pour in the stout. Sam was soon known across Yorkshire as the beer-drinking bull and his story even made the local papers.

In another of the photographs in the restaurant, a young boy wearing homemade jodhpurs is proudly striding along next to Charlie and Sam at a show in Derbyshire. He's Roger Nicholson, Charlie's young son, and he looks as pleased as punch to be leading the parade with the family's winning bull.

Nobody then could have imagined that, more than six decades later, Sam the beer-drinking bull would still be showcasing the talents of Cannon Hall Farm, and that the cheeky little boy in the photograph would have turned a rural homestead in Yorkshire into one of the most successful open farms in the country.

In 2019, in recognition of Roger's 60 years at Cannon Hall Farm, his sons Richard, Robert and David presented him with an anniversary herd of Shorthorn heifers – plus a big impressive bull called Jeremy. It's just one chapter in what continues to be a remarkable story for the Nicholson family at Cannon Hall Farm, a story now shared with television audiences and social media worldwide.

And all because of a little white bull…

Bull goes on the BEER

FOCKERBY RING LEADER, a four-year-old pedigree bull, gets an extra reward whenever he wins a show prize these days—A PINT OF STOUT.

Last time he won was at the Eckington Show, Derbyshire, and afterwards his owner, Farmer C. B. Nicholson of Bank End Farm, Worsbro' Dale, near Barnsley, with the "prize card" tucked under his arm, led Fockerby off to the refreshment tent for his pint.

"He gets his drink every time he wins a prize," said Mr. Nicholson.

"It used to be beer, but early this year Fockerby developed a taste for stout, and now he will not drink anything but that."

Mr Nicholson bought the bull for 55 guineas when it was ten months old.

It first displayed its taste for beer two years ago, when Mr. Nicholson offered it a drink as a reward for a win at Barnsley Show.

Since then the animal has become a hardened drinker—so far this year it has gained three firsts and two seconds on its five appearances in the show ring.

The doctors have told me I shall be out of hospital soon. This accident will not affect my nerve. I shall go back to my work on the bridge.

Chas. Radford.

Needle and thread

A newspaper cutting is a reminder of Charlie Nicholson's beer-drinking bull

Charlie with Sam the bull

1

Roger the hero

There was no one in the farmyard. As usual, Roger had raced home from school as soon as the bell rang at four o'clock. At 13 he wasn't expected to do the dawn milking, but his dad Charlie had put him in charge of the afternoon shift, and Roger was determined to show him that he was a proper farmer in the making. He ran into the house to say hello to his dad and mum Rene, threw off his coat and changed out of his school uniform before heading to the cow shed.

Perching on the little wooden three-legged milking stool, Roger pressed his head firmly into the side of Polly the cow. If there was the merest twitch from her haunches that she was about to kick, Roger was primed to get out of the way smartish. He'd been in charge of afternoon milking from the age of eight, and while the Nicholsons' cows Polly and Molly were getting older and were unlikely to kick whoever was milking them, he still wasn't about to take any chances. They weren't going to win any can-can competitions, but Polly and Molly could still kick like billy-o.

Sitting in the draughty cow shed that chilly day in October 1956, summer felt a long time ago. All the crops had been harvested and the fields were being prepared for next year's planting; it was all part of the cycle of the year at Bank End Farm in Worsbrough Dale, South Yorkshire. As autumn eased into winter, young Roger was glad of the warmth of the cow's side against his shoulder as he rhythmically milked the trusty old Shorthorns and daydreamed about one day playing for Barnsley FC.

The pail quickly filled, so Roger finished up the milking and poured the steaming creamy liquid into the milk churn, before making sure it was sealed nice and tightly in readiness for the milk round the following morning. At weekends he'd help his father doing the rounds, but Monday to Friday it was all about school. 'More's the pity', Roger often thought…

All the animals had been fed and settled for the evening and everyone had gone home, so Roger decided to do a bit of ratting before tea. He picked up his trusty 410 shotgun and headed to the corn store. As every farmer knows, your farm can be as clean as a whistle, but where there's pig muck and stores of corn, rats inevitably follow. So Roger was doing the family a favour keeping the rodent population down – even if, strictly speaking at his age, he shouldn't have had access to a gun.

But it was what happened next which made him a family hero…

As Roger headed towards the grain store where the rats would gather, the unmistakable smell of smoke hit his nostrils and his heart started to beat faster. For any farmer, a fire on the farm can be devastating – destroying livelihoods in a matter of minutes.

The dark evening air felt thick in his chest as he rounded the corner to see a huge blaze where the stacks of corn were stored. The wind was up that night, blowing the smoke towards the barn where the animals were kept. There was no time to call for any help as the animals would suffocate if the smoke got to them. Thick plumes were billowing from the base of the stack and Roger could hear the animals grunting and lowing in fear. There were eight piglets and three young calves in there. He had to get them out.

With lightening-quick thinking, Roger had decided that the calves should be moved first and the piglets second as the smoke would be less toxic at ground level. As the flames licked the ground in the stockyard, Roger ran into the barn making a beeline for one of the terrified little calves that was twitching in distress. He hauled it up into his arms and swiftly carried it to the empty bull shed across the yard away from the fire. He then turned on his heels and fetched the second one, then the third, and then headed back in to save the pigs that were squealing and charging around the barn, desperately looking for a way out.

It was difficult not to trip over the panicking piglets as the smoke got thicker, but adrenalin surged inside him and he was determined to save every last one. Each time he placed an animal down in the empty shed, he'd take a big gulp of air and battle on back in.

Had he got them all? The animals were running around in confusion and fear and Roger had lost count. If there were any left in there, surely it would be too late?

The Bantams in the hen house got wind of the disturbances and added their voices to the cacophony in the yard. The farmhouse was a fair distance away from where the fire was blazing so Roger's parents were oblivious to the drama, but thankfully a passerby had seen the flames and called the fire brigade. And while Roger was saving the animals, two fire wagons were racing over from nearby Barnsley with sirens clanging.

It was only when Roger counted and recounted the animals and knew they were safe that he could turn his attention to the fire itself. If the flames got hold of the buildings then it would only be a matter of time before everything was destroyed. Suddenly he heard the noise of the fire engines approaching the farm – it was the sweetest sound he'd ever heard. Finally he was able to breathe and take stock of the scene before him.

Seconds later, as the fire relentlessly pulsed with heat, the fire engines had barely come to a stop before the brigade sprang into action like worker bees. Within seconds gallons and gallons of water were swamping the blaze.

Finally, alerted by the fire engines, Roger's parents ran out of the farmhouse and were shocked to the core when they saw the fire and the expression on their son's face. It was obvious he'd been in the eye of the storm and Rene ran to hug Roger as it dawned on Charlie that his son had not only been single-handedly dealing with the fire, but he had also saved the animals.

The fire brigade continued to quell the blaze and worked through the night to make sure every inch of the fire was damped down, while the Nicholsons learned that two other local farms had been victims of arson attacks that night. Charlie scolded Roger for not reporting the fire to them straight away as he put himself in so much danger, but recognised his son was indeed a true farmer for saving the animals first.

They all understood how close they had been to losing everything.

As they collectively breathed a sigh of relief, Roger went back to check on the animals, then dropped the latch on the outbuilding and headed indoors for the night.

The ratting could wait until tomorrow.

Growing up, Roger had always been adored by his parents and three big sisters. Olive was born in 1924, followed by Shirley in 1930 and Beryl in 1932. Charlie and Rene's first longed-for boy, who they named Alan, was born between Olive and Shirley, but died on the night of his second birthday. He'd been in hospital for what everyone had believed to be a minor operation – it certainly would be in modern times – and when Charlie and Rene had brought him back home from hospital everything had seemed fine. But that night little Alan just slipped away. The fact the little tot died on his birthday somehow made the tragedy even harder for everyone to bear. Celebrating as the day began and heartbroken by the end of it.

Devastated by their loss, when Roger eventually came along in 1943, Charlie and Rene were almost too protective, watching over him like a hawk and always trying to second-guess any potential dangers. His sister Beryl, 11 years older than Roger, recalls sitting by his bedside, night after night, making sure he was safe until he fell asleep, watching him breathe and dozing off beside him.

But as Roger grew into a toddler, the plucky little rascal wasn't to be contained by his anxious family. He'd charge around the farmyard and loved to watch his dad out at work, tending the cows and sheep or taking out the shire horses to pull the ploughs in the corn fields. Every now and then, Charlie would scoop Roger into his arms and hug him tight. "This is Bank End Farm, son," he'd tell him. "One day all this will be yours."

Early days at Bank End Farm

In 1830, the MP William Cobbett, describing the industrialisa-
tion of rural Yorkshire, wrote: "All the way along from Leeds
to Sheffield it is coal and iron, and iron and coal." He obviously
didn't visit the village of Worsbrough Dale, around three miles away
from Barnsley. The village, whose name means "Weorc's fortified
place" was listed as being in "the wapentake of Staincross in the
West Riding of Yorkshire" in the 1086 Domesday Book.

Worsbrough Dale has always had a thriving community with
shops, schools, pubs, churches and even its own brass band. In the
churchyard of the Grade One listed St Mary's parish church, there
are several gravestones belonging to ancestors of the Nicholson
family, including that of John Nicholson (1769-1827), who originally
bought Bank End Farm in Worsbrough Dale in 1798.

After John Nicholson died, Bank End Farm was passed down
to his brother Samuel and on his death 20 years later, the farm
then went to the youngest brother, William. William died two years
later in 1849 and his wife Frances held the deeds of the farm until
her death in 1855. It was handed down to William and Frances's
son Charles, and on his death in 1876 his wife Mary owned the
farm until she passed away in 1910. Charles and Mary's son, Joseph

Samuel, was the next in line, finally passing it on to his son Charlie, Roger's father, in 1937.

Roger was born on 8 March 1943 in Bank End House, or the "new house" as the family called it. The original farmhouse at Bank End Farm, where his three elder sisters were born and generations of Nicholsons had lived before, was built in 1652, so Bank End House was relatively modern in comparison.

The handsome detached house had a large vegetable garden where Roger's mother grew fruit and vegetables to feed the family, and there was always plenty of meat from the farm, not to mention the pheasants, partridges, grouse, hares and rabbits that Charlie would shoot. The family wanted for nothing.

Although the farm had been passed down through the Nicholson wives as well as sons, the business was very much regarded as a male concern. Growing up, none of Charlie and Rene's daughters entertained the idea of taking on Bank End Farm: Olive, having left home as a teenager to join the RAF, had married an engineer called Vincent Wilson; Shirley got a job as a builder's secretary before marrying local farmer Laurie Mount; Beryl also married a farmer, Brian Robertshaw, but the couple gave up farming after eight years and took on a village newsagent.

Beryl is Roger's only surviving sister. As a child during the Thirties, she remembers being very timid and hiding behind her mother's skirts, never saying boo to a goose. She enjoyed growing up on the farm but was scared of the animals and would always make a point of avoiding walking between any of the cows. "What made things worse," she recalls now, "was the farm hands knew I was a scaredy-cat and when Shirley and I used to bring them their lunch in the fields, they'd be armed with mice, ready to pelt us. It's amazing how fast you can run when you see a mouse flying towards you!

"Even so, it was a lovely place to grow up. So much space! And I'll always remember the celebrations we used to have. When Roger was born my father threw three parties! And they were proper 'sit-down-to-a-meal' dos. There was one for the family, one for the staff and tenants at the farm, and one for friends, so my poor mother was kept very busy. And to think, she'd just given birth!"

With very little in the way of labour-saving devices, it was all hands to the pump – or rather, that is, all female hands to the pump – helping with the daily chores such as beating the rugs, airing the feather mattresses, darning socks and doing the laundry. Beryl recalls Charlie buying her mother a huge green washing machine, which would be loaded up every Monday. "It sounds modern, but it did little more than swish the clothes around. Everything still needed to be rinsed by hand and wrung through the mangle."

Back then, Monday was washing day – the air was always cleaner at the start of the week because the factories in nearby Barnsley didn't operate on Sundays. Tuesday, meanwhile, was the day when it was Beryl's job to tackle the ironing.

The oven, Beryl remembers, was heated by a coal fire. "We'd push more coals underneath the oven depending on whether you wanted it hot or very hot," she says. "I don't know how Mum managed to gauge how long to cook things, but she made everything from scratch. We never had to buy anything from the grocery. She'd bake bread, make jam, pickle vegetables and do whatever was needed."

With no fridge, all the perishables were kept downstairs from the kitchen in a cold stone larder and a ventilated meat safe. "During the war, farmers were allowed to kill a pig for their family every so often, so we'd make pies and sausages and all sorts," Beryl remembers. "We certainly never went hungry and Dad was very generous and would make sure that anyone else who needed food got it."

If there was anybody who needed a helping hand, Charlie was always there for them. "From time to time he'd say to us, 'Have you any toys you've finished playing with?' and he'd give them to families at Christmas who couldn't afford to buy their children anything."

And when Charlie joined the Barnsley Rotary Club in the Forties, he'd always offer to host overseas visitors, happy to show off his bit of the beautiful Yorkshire countryside.

Roger recalls how his dad mentored a local lad called John Linford, who was born to be a farmer. John was a frequent visitor to the farm and after leaving junior school he was one of the first pupils at White Cross Secondary School, which was built on land that was previously part of Bank End Farm. Being a bright lad he could have had his pick of better schools, but he liked White Cross because it was handily placed for him to call on the farm on his way home from school. "John always thought very highly of my father and credited him for teaching him the right way to have a successful life and career."

All in all it was a happy home, a great environment and the four children thrived. Charlie and Rene always strived to do the very best for their family, tackling the challenges of bringing up four intelligent individuals who would all go on to make their parents proud.

Roger clings on to Blossom the shire horse

3

Roger's childhood

When he was a little boy, Roger could never get enough of farming, unlike his sisters, who were generally kept busy indoors helping their mother. As soon as he was able, little Roger would follow his dad and the farm workers to the fields, climb up onto the rickety iron harness of one of the family's shire horses, and cling tightly to the reins. Blossom, the magnificent mare, would steadily rake the newly cut hay into lines to be collected and Roger would beam from ear to ear.

The rig's frightening looking metal tines didn't bother the young lad as it combed the crops. Roger just concentrated on holding on, keeping the lines straight and showing his father he was a natural. When you're eight years old and you want to grow up to be a big strong farmer like your dad, your legs can't grow quickly enough, Roger remembers. "They obviously thought I was all right at it. It would take a good couple of hours to do each field and they'd just leave me to it!"

Roger loved Blossom. If the shire horse was spruced up and looking her best for horse-and-trap duty, he couldn't resist clambering up to her saddle and being king of the castle. The fact his little legs didn't reach the stirrups was neither here nor there. The family treasures a photo of Roger aged about three years old, wearing his trademark cap which he liked to doff to all the ladies he'd meet. He's sitting in Blossom's saddle and couldn't look happier.

"Blossom was such a fantastic horse," he says. "She'd be out pulling a plough on a field nearby and then, when it got to four o'clock in the afternoon, she somehow always knew it was time to come home and would stop for the day. I think she must have had some kind of built-in alarm clock."

At any given opportunity, the youngest Nicholson would shadow his father around the farm. He soon learned how to cradle a baby spring lamb in his arms and hand-feed it if it was struggling to suckle from its mother. He'd snort at the piglets and practise counting them up as they lay there contentedly feeding, and he spent many happy hours chasing his pet hen around the yard. It's little surprise that, for as long as he can remember, Roger never entertained ambitions to do anything else but be a farmer. "I didn't mind school too much, but I don't think there was ever a time when I thought I'd do anything else when I grew up," he says.

There were around 60 children in Roger's year at the local village primary school, Worsbrough Dale. Coming from the farm rather than the smaller houses nearby, Roger initially got a bit of stick from the other kids at school. But what young Roger lacked in height, he more than made up for in personality, and he quickly earned their respect and became one of the gang.

Together with his schoolboy chums Freddie, Billy and Michael, Roger made the most of living in the countryside. The lads, who lived on the local estate, would hop over the Nicholsons' farm gate and call for Roger to come out to play, and they would all head off together on an adventure, loving the acres of space they had to explore and hone their cricket skills. Roger didn't have a care in the world.

Yet despite having plenty of land to farm, making a living from it wasn't always easy for the Nicholsons. The family always got by, but that meant Charlie and Rene had to work their fingers to the bone. Land needs to be maintained and livestock has to be cared for every single day of the year, in all weather conditions. There's no such thing as a free weekend; even on Christmas Day, there's work to be done.

Yorkshire winters can test the mettle of even the toughest farmer. When torrential rain and heavy snow falls, tracks can turn to black ice and the biting winds howl and numb your fingers and toes. It takes a special kind of person to keep going when there's no let-up in the harsh winter weather. Farmers are on call twenty-four-seven during lambing season, and if the weather isn't on your side a whole year's worth of crops can be destroyed in a matter of days. Farming is a vocation. If you can't hear the calling, it's not for you...

The saying goes that where there is livestock, there is deadstock, and it is not a given that a flock of sheep, a litter of pigs or a herd of cows will make you money. Breaking even is sometimes the very best you can hope for.

Ever the optimist and by all accounts a charismatic businessman, Charlie tried several ways to make the farm more profitable for his family. "There was many a time that Dad would be up and out at four o'clock in the morning selling buckets of potatoes to the local miners," Beryl recalls. "He'd call out '28lbs for a bob...' Then there would be turkeys and geese at Christmas, harvesting, threshing... There was always something. It was always busy."

Although the family owned the farm itself, Charlie rented a

further 150 acres of land from a local potato merchant. He used the land in various ways throughout their time there, sometimes growing crops, other times pasturing animals. "I suppose Dad was a bit of a wheeler-dealer, really," says Roger. "In my lifetime at Bank End Farm, he did a bit of everything. When we got our first washing machine, for instance, he bought a few others in order to sell them on. He'd also buy and sell property, as it wasn't expensive in those days. You could probably buy a whole row of houses for 100 quid!"

Coming from a long line of famers at Bank End in Worsbrough Dale, Charlie was highly respected in agricultural circles. During the Second World War, he was offered a new opportunity working for the Ministry of Agriculture. Even if he had been the right age to enlist – he had been too young for the First World War and was too old for the Second World War – farmers were often exempted from service in order to ensure that the nation was fed and nourished during the conflict.

The Ministry of Food became the sole buyer and importer of food. They regulated prices and guaranteed that farmers had a market for their produce, and Charlie was recruited to advise farmers as to what to grow depending on what was needed at the time.

With Charlie working for the Ministry of Agriculture and the girls still at school, the war years weren't a time of terrible hardship for the family, unlike many families around Britain. While coal was rationed, they always had plenty of firewood from their local wood, so life ticked along nicely. Evenings were filled listening to the radio and playing Ludo and Snakes and Ladders.

During the height of the conflict, rural areas of England were certainly safer places to be than major cities. It could have been a very different story if the Nicholsons had lived just half an hour away in Sheffield. Over the nights of the 12th and 15th December 1940, more than 78,000 homes were damaged and around 40,000 people were made homeless in the Sheffield Blitz.

At the end of the war, Charlie received a letter from the Right Honourable Robert Hudson, the Minister of Agriculture and Fisheries, thanking him for his work during the war. It was dated 8 May 1945 – VE Day – and read: "On behalf of his Majesty's Government, I wish to thank you for the service you have rendered to the nation during the war. The task of British agriculture, an arduous, indeed a vital one, was to keep the nation fed. With your help it has been done." It concluded: "I am confident you will always be proud of having played so important a part in the contribution which British agriculture has made to our Victory."

In 1953, with rationing almost at an end, the Government approved a scheme to subsidise farmers to help them produce more meat. In his new role as a Calf Certifying Officer, Charlie would visit farmers who had applied for grants and decide which animals made the grade. Farmers were rewarded for each calf that was raised to the right standard. A purebred Friesian calf, for example, wouldn't qualify for a subsidy as it was a dairy cow and the priority at the time was rearing beef cattle; Shorthorn cattle, which the Nicholsons traditionally kept, were recognised as dual-purpose dairy and meat animals; Hereford and Aberdeen Angus cattle were solely bred for beef.

Farmers would sometimes try and pass off dairy cows as beef cattle, blackening up their legs with boot polish to disguise them.

But canny Charlie never fell for ruses like that, although he could understand why farmers would go to any lengths to get by as it was such a tough industry to survive in.

Roger recalls going along with his father when he made his certification checks, rounding up the cattle for him to check over: "It was a great education to see him in action and I loved having that time with him to myself."

Always driven by making a better life for his family, it was during the 1950s that Charlie started to make headlines with his beloved pedigree shorthorn bull Sam, aka Fockerby Ringleader. Charlie had a nose for a champion and had seen the potential in the 10-month-old, paying the then substantial sum of 55 guineas for him.

His gamble paid off when Sam went on to win numerous county shows around the north of England. "They weren't always the creme de la creme of county shows," Roger explains. "Just local shows that would happen most weekends during the summer months. Dad would go to between 15 and 20 a year, and in the four years he showed Sam, the bull always did him proud. On one occasion, Dad took a black and white Friesian bull to show instead, but as he was tying him up, the bull tried to mount him. I think that was the end of that particular bull's county show career…"

Little Roger also got the bug for competing at shows, taking along his little bay pony Rosie to gymkhanas. But he didn't have his father's winning streak. "I usually came second last," he laughs. "I was always a bit on the small side, but I loved that pony and she was really smart. My friend Malcolm used to ride her in the musical chairs competition as he had longer legs than me and could dismount and get to the chairs first. He'd always win!"

Meanwhile, back at home, Rene worked hard bringing up the

children. "Because we lived on a farm, we were luckier than most," says Roger. "But if my mother had nothing for dinner, she'd get her air gun, shoot a pigeon, pluck it, gut it and then bake it in a pie. She was pretty good with that gun…"

When they each turned 11, Shirley and Beryl both attended the nearby Beech Grove independent school, while Roger headed to the local boys' grammar school, Barnsley Holgate. "I quickly learned that once you go to a school where everyone's passed their 11-plus, no matter how good you were at your junior school, you can never be as clever as the cleverest."

As well as being academically lauded, the school had a fantastic reputation for sports and young Roger was determined to excel. He was still on the short side, but that didn't stop him running his legs off in the cross-country team. "I wasn't really built for it, but I used to grit my teeth and run so hard that I was often sick as a dog when I crossed the finishing line. But you have to give it a go, don't you?"

Roger also played rugby and cricket, and he always loved athletics. But as much as he tried, he never made it past the trials to join the football team. The standard was very high and five of the school's players were already playing for the Yorkshire boys' squad. Two were in the England team. And Brian and Jimmy Greenhoff, two of Holgate's other ex-pupils would later go on to play football professionally for Manchester United. Other famous alumni from Barnsley Holgate include the television personality Sir Michael Parkinson and the legendary cricket umpire Dickie Bird OBE.

One of the highlights of the sporting calendar was the boxing championship, held in the school's very own boxing ring. Roger is still very proud of his prowess, although he still smarts as he recalls one particular fight. "I only ever lost one bout and that was only because the other kid were so much bigger than me. I was robbed! The fights were always very exciting, though, and we'd have a good turnout in the audience. But my father was always too anxious to watch and would only come back in at the final bell!"

Getting good reports in English, history, geography and arithmetic, Roger was making the most of grammar school life, yet in his heart of hearts he always knew that he would go into farming as soon as his school days were over.

Just as life seemed to be on an even keel for the Nicholsons, in 1953 the family received some devastating news. They were informed that Bank End Farm was subject to a compulsory purchase order from the local council.

Councils had been granted such powers by Parliament for the 'public benefit', with little consideration given to the impact it had on the individuals who would be having their land or property taken from them. In this particular case, following the war and the subsequent growth in population, there was a shortage of housing stock in the area, and the council had earmarked the land that the Nicholsons were farming at Bank End for new homes and playing fields.

The family owned only a small proportion of the land that they had lived and worked on for generations; the majority of it was rented. The compensation that Charlie would receive – a meagre £60 an acre – was determined by the agricultural value of the land and gave no consideration to the value of the buildings built on it. The fact that there was a working business was neither here nor there.

With the rural landscape going through a period of rapid change in the years after the Second World War, compulsory purchase orders weren't unusual and the Nicholsons had already been forced to give up various chunks of their land. On those previous occasions, they had still managed to make a living with less acreage, but this time, the council's plans were life changing. When a small fraction of land is taken, it's one thing, but this time it would completely destroy the Nicholsons' home and livelihood.

With no space to rear animals or grow crops, the farm simply wasn't viable. Although Charlie was managing to keep his head above water thanks to his work for the Ministry of Agriculture, fate seemed to be conspiring against him and his family having any sort of farming future at Bank End.

"It was a devastating period of all our lives, but especially for my father," says Roger. "He thought that he would live out the end of his days there and, suddenly, that whole future he'd mapped out in his mind was torn apart. Those acres of fields that he had ploughed, nurtured and harvested for all those years would disappear from the landscape and his dreams of my future at the helm were also dashed."

It took a long time for the family to come to terms with the fact that they would have to leave Bank End Farm and start again somewhere new, but they had no choice. Retiring wasn't an option – and in any case, Charlie didn't want to give up farming. It had been his life and he wanted it to be Roger's life, too.

With savings from his Ministry of Agriculture job and the compensation he received from the compulsory purchase scheme, he began to set his sights on new horizons…

4

Daring to dream

As you arrive at the foothills of the Pennines just outside Barnsley, the landscape opens up like a magical storybook, with a patchwork of fields stitched together with traditional dry-stone walls. Clutches of trees, farms and pretty villages pepper the hillsides, and it feels like a world away from the busy towns and cities nearby.

Whatever season you choose to visit, be it in the depths of winter when snow blankets the rolling hills, the cool, bright spring when the trees are heavy in blossom and new lambs gambol in the fields, or the summer and autumn when light and colour paints the countryside in lush new hues, Yorkshire is ever-changing but always beautiful. No wonder locals call it God's Own Country.

If you happen to be headed towards the picturesque village of Cawthorne, you can't fail to be wowed by the stunning Georgian country house Cannon Hall, with its breath-taking facade. Sitting squarely in 70 acres of beautiful national parkland, the house was once owned by the Spencer-Stanhope family, who made their fortune in the local iron industry.

John Spencer, a Welsh hay-rake maker, bought the estate in 1660 and the house was subsequently extended in the 18th Century by the celebrated architect John Carr. The manor was then handed down through generations of the Spencer-Stanhope family before the estate was broken up and sold to Barnsley Council in 1951.

Nearby in the village of Cawthorne, the Spencer Arms pub keeps the family name alive. The village is picture-postcard pretty, with handsome sandstone houses and beautifully kept gardens. There's a village hall, grocery store and tearoom, two churches, a post office and even a museum filled with curiosities. Want to see a man's boot that was struck by lightning, or a two-headed lamb? This is the place to visit.

Parts of the village remain unchanged from the days when the Spencer-Stanhope family lived at Cannon Hall, but by the time the country house came to be sold in 1951, great swathes of its land had been destroyed by open cast mining. Traditional pits had become uneconomic to run, so shallow open cast mining techniques were introduced, meaning parkland that had previously been enjoyed as a beauty spot was chewed up by diggers. Even the property's stunning landscaped gardens were ruined.

With Cannon Hall looking a shadow of its former self, Barnsley Council purchased the house and appointed a curator to create a collection of the mansion's artefacts. Six years later, in 1957, it reopened as a museum.

While the house and beautiful gardens at Cannon Hall were being brought back to life, its adjoining farm, farm buildings and extensive farmland were put up for auction – and that's where the Nicolson family comes back into the story.

"Cannon Hall Farm was probably the premier address in the area," Roger explains. "My dad had often visited this side of Barnsley and knew the area well. So when he heard the news that the farm was up for sale, he saw the opportunity of taking it over. It wasn't a huge farm in the grand scheme of things and would be workable enough providing we had some extra help. Plus he'd always farmed with a view to passing it on me and this way the farming legacy could continue."

So, on Wednesday 10 April 1958, a suited-and-booted Charlie Nicholson headed to the Royal Hotel in Church Street, Barnsley, where at three o'clock that afternoon the family's potential new home was going under the hammer. Charlie intended to bid on Cannon Hall Farm, described in the particulars as "a first-class mixed farm extending to approximately 126 acres".

Making a profit from farming was never guaranteed, and the family understood only too well the risk of failure, but a new start was just what they needed. Charlie had visited the farm many times with Roger and could picture them working the land together and giving the property a new lease of life.

Set around a pretty country courtyard at the back of Cannon Hall, the farm buildings that were up for sale along with the land included a solidly built three-bedroomed brick house and several other buildings, including a garage and workshop, a grain store, stables, a grinding chamber for crops and a cow shed. There were also two timber Dutch barns, a place to keep pigs and storerooms for coal, potatoes and farming equipment.

As if that wasn't already enough, as well as more than 125 acres of farmland, there was also the beautiful stone Tower Cottage. Looking like something out of a fairy tale, the Grade Two listed building was a miniature castle complete with a square turret. It was built in the late 18th Century as a Gothic folly and at one stage during the Spencer-Stanhope years served as the home of the

Cannon Hall Estate gamekeeper. It certainly wasn't the usual kind of outbuilding you'd find on a farm.

On the day of the auction, Roger, then aged 16, had gone off to school as normal, knowing that by the end of the day Cannon Hall Farm could, potentially, be theirs. "In our heads, we'd practically moved in already."

To all intents and purposes, it was a normal working day in Barnsley, but there was a real buzz at the Royal Hotel as the auction room hushed for Lot One. As the description of Cannon Hall Farm was delivered to the bidders, Charlie felt confident that he was in with a chance.

But before he could really take in what was happening, bids started coming in thick and fast. The auction seemed to be running away from him. Charlie had been determined to hold his nerve, but as the numbers increased, he could feel his heart beat faster and his palms started to sweat.

£4,000... £5,000... £6,000...

Charlie's head was telling him it was a hopeless cause, but his heart wanted to stay in the game. He had kept silent throughout the auction and it was only when the hammer was about to go down on the sum of £7,000 that he finally got to his feet and said, "£7,100". To his relief, the extra £100 sealed the deal.

In Charlie's heart of hearts, he knew he had paid more than he could really afford. He certainly hadn't intended to spend that much, but perhaps a certain stubborn streak drove him to break his own rules.

He might have secured the shell of his new home, but he still needed to purchase the animals and certain farm equipment to make the property work for the family. But he had every confidence in himself. Never one to rest on his laurels, he would do whatever it took to make sure that Cannon Hall Farm was a success.

"I remember that day so vividly," says Roger. "As soon as I got out of school, I rushed home to find out if Dad had bought the farm. When he saw me, he tried to pretend he'd failed, but that twinkle in his eye wasn't fooling anyone. He calmly shook my hand and said, 'We got it, son'."

5

New life lessons

In order to get things ready to move in, Charlie took Roger along with him to the official farm sale at Cawthorne where the contents of the farm were to be auctioned. They could bid on everything from hammers and nails to farm animals, tractors, cultivators and ploughs. Coming from a farm, the family already had lots of farm equipment, of course, but when the things that you are going to need are already there in situ – like an enormous corn grinder, for instance – you may as well add it the shopping list. For Roger, it was another part of his farming education.

There were lessons to be learned, but unfortunately mistakes to be made, too. "My father had bought a newly calved heifer in the sale," Roger explains. "Then one of our new neighbours mentioned to me how much he had wanted to buy it and asked if I would sell it to him. Knowing it would be good to keep the neighbours sweet and not really thinking it would be a problem – we'd bought plenty of other livestock, after all – I agreed to sell it on to him for £60, which is what I thought my father had paid. Only it wasn't, it was £70. We'd only just moved there and already I was making stupid decisions. I felt really bad about it. Luckily, the buyer found out I'd made a mistake and honoured the extra £10, so nobody lost out.

But nobody won either. I just lost a bit of pride. I still remember that mistake to this day."

At 16, Roger was still at school, but really wanted to prove he had the maturity of a working man. With farming blood coursing through his veins he felt every inch a farmer, but he still had a long way to go. Luckily it was a supportive farming community and whether buying or selling, nobody ever tried to rip anyone else off, it just wasn't done in those days, Roger remembers. Fair bids were offered in exchange for goods and services, and soon the farm was equipped with everything needed to make a fresh new start.

And yet, having initially been confident about the family's future at Cannon Hall, Charlie began to have second thoughts about what he'd taken on. The realisation of the monumental task ahead made him seriously consider selling on Cannon Hall Farm before they had even planted their first crop. Even Roger's school history master commented on the small fortune Charlie had shelled out, intimating he might have made a big mistake. "We knew it was going to be hard," says Roger. "But I was always determined that we should give it our best shot."

For six months before the Nicholsons moved in properly, a family friend Arthur Fisher and his wife moved into Cannon Hall Farm to keep things ticking over while Charlie and Rene sorted out everything in Worsbrough Dale. The Nicholsons had been there for generations, so it wasn't just a matter of packing up overnight and moving on.

Roger often went to stay with the Fishers at Cannon Hall Farm, savouring the rice pudding that Mrs Fisher made daily and helping out in any way he could. "It was a new adventure for me. I'd never had that sort of area to roam around before and the farm was so full of history. The barns seemed vast to me and I'd walk around and imagine the different animals we could keep there. Everywhere I looked I could almost see and hear people living and working there. In one of the granaries there was an old snooker table that had been left by an old tenant – every barn seemed to have a story and it was a dream come true that the place was ours."

During Roger's frequent stays at his new address, Arthur passed on his wealth of farming knowledge, knowing that Roger would one day be running Cannon Hall Farm himself. "Arthur was a so and so for cleanliness," Roger recalls. "I remember he used to keep fancy pheasants in pens around the back of his house and he was immaculately clean himself. In fact, one of the first things he instilled in me was how important it was to keep the farm clean in order to prevent disease. He'd been a cattle dealer and he taught me all about the different breeds of cows, the ailments they might get and general tips on animal husbandry to stand me in good stead. The lessons Mr Fisher taught me were so invaluable and I'm still so grateful to him."

"I learned quite a lot about the veterinary side, too. Farmers couldn't administer penicillin themselves for animals in those days; it had to be done by a vet. And if you called in the vet and the animal survived, the animal would be worth more than the vet's bill. Now, of course, the vet's bill can be more than the animal is worth."

As a child, Charlie had suffered from rheumatic fever and by the time he was 50 he had begun to have heart problems. When the Nicholsons moved into the farm, extra hands were taken on as Charlie's health was going noticeably downhill. Although Charlie hadn't had a heart attack, he had blackouts and the doctor ordered him to stay in bed and rest up for a few weeks. Nowadays a heart bypass would be a fairly routine operation, but in the middle of the 1900s it was a very different story.

Whether he wanted to or not, Charlie was forced to slow down and some farm work was always going to be completely out of the question as he was unable to walk up the steep hill back to the farmhouse from the lower fields. Luckily, he had his strong son Roger by his side. "I took on as much as I could, but I could see Dad's health wasn't good. He was really struggling and no matter how much Mum and I nagged him, he still did more than he should."

Back in Worsborough, the family had bought a Friesian cow with the idea of starting a dairy at Cannon Hall Farm. The cow's pedigree name was Bulcliff Gwyneth and cost the huge sum of 140 guineas at market. Unfortunately, she never realised the potential the family had hoped for, so the plans to make Cannon Hall Farm a dairy farm had to be shelved fairly quickly. Instead, to make sure there would be money coming in straight away, the farm workers immediately planted around 10 acres of potatoes and bought in some calves to rear and lambs to fatten for market.

6

Family heartbreak

Less than a year after moving into Cannon Hall Farm, Charlie died. Roger takes up the story: "On the morning of 21 April 1959 someone had ordered a ton of potatoes and Dad had taken it upon himself to riddle the whole lot on his own. It's such a physical job removing all the loose mud and sieving out the tiny spuds, but he wanted to be at the Rotary Club by lunchtime, so he didn't want to hang about.

"After lunch at the Club Dad came back home and went to fetch some hay bales down from the hayloft to feed the cows. It was then that he just dropped dead. He was only 56."

That afternoon, when Roger came back from school, he stepped off the bus and one of their farmers, Eric Roe, was waiting there to give him the terrible news of his father's death. As soon as he saw the expression on Eric's face, Roger feared the worst. Devastated, Roger wanted to go up to the hayloft to see his father's body, but Eric pulled him back. "He told me there was nothing I could do," Roger recalls. "He said it was more important that I comforted Mum."

Roger remembers his mother being incredibly strong when Charlie died. Rene had always been a hard worker and, although she was obviously sad about losing her husband, she showed remarkable resilience and determination to carry on for her son's sake. Having endured the heartbreak of losing her first son, she was comforted that Charlie had lived a fulfilled life, and the fact that he ended his days at Cannon Hall Farm was a fitting end to his endeavours. "My mother was a remarkable woman," says Roger. "She was fiercely determined and independent."

Rene and the girls organised the funeral and, in the days and weeks that followed, Roger steadily began to face up to his new responsibilities. "The day that Dad died, a family friend called Will Roe – Eric's brother – promised that whatever he could do to help me, he would. And he continued to do just that until the day that he retired in the Seventies."

Roger received a letter from the Ministry of Agriculture advising him on the steps he needed to take in order for the farm to operate without his father and he also had many offers of help through Charlie's connections at the Rotary Club. Roger may have been the man of the household now, but he could take some solace in the knowledge that he wasn't expected to steer the ship alone.

On the day of Charlie's funeral, the village church of St Mary's, back in Worsbrough Dale, was packed to the rafters with the Nicholsons' friends, family and Charlie's work associates. Charlie had been such a charismatic and popular man, that it felt as if the whole of Barnsley's great and good had turned out to pay their respects.

In their notice of the funeral, the *Barnsley Chronicle* carried a photograph of a very smart Charlie Nicholson at a Rotary Club event. The accompanying article spoke of the many fellow Rotarians and farmers from Barnsley and several surrounding districts who had attended the funeral. It listed some of his achievements: "Mr Nicholson was a member of the Barnsley branch of the National Farmers Union for many years and also of the Ploughing Association at Worsbrough and Wentworth. He was also the last chairman of the Barnsley Agricultural Show."

Roger is hazy about the details of the day, but he remembers he took his father's death badly and was very upset throughout the service. His mother, as always, was remarkably stoic. "I really felt like Dad was there with us when we were in the church," Roger says. "But then, at the end of the service, I remember that feeling like he was gone. It was as though a line had been drawn and it was time for us all to move on."

And so, the morning after his father's funeral, it was back to work as normal for the farm workers. Roger's sisters returned to their own homes and a new chapter in Roger's life began.

From that moment onwards, Roger's days at school were numbered. He only had a couple of months to go until he was due to take his O levels anyway, and he ended up sitting a handful of exams rather than the full academic list. The teachers knew that the farm was going to be Roger's future and decided between them that he didn't need to sit exams in any subjects he'd have no need of. "I did pass a few," he says. "But it were a poor effort for the standard of that school. I knew that I didn't need any qualifications on paper for what I was going to do. Had my dad lived, I probably would have gone on to agricultural college, but obviously that was impossible. As it was, I was going to get all the on-the-job training a farmer could need."

From the age of around 20, Rene's hearing had been deteriorating and although she now lived in a world of silence, Roger had learned how to communicate with her with their own version of sign language and he was an invaluable support to her in the aftermath of Charlie's death. For her part, Rene took on the bookkeeping and helped Roger with the day-to-day tasks of the farm. After all, she had spent enough time watching Charlie running the business to know what to do. According to Charlie's will, the farm would be in her name until Roger's 21st birthday, and she was determined to help make it viable for him.

Even though Roger had older sisters, it had always been agreed that the farm would be passed down to him.

Once Roger's school days were over, taking over the farm meant a different kind of education for the teenager. "I can't pretend it came naturally at all," says Roger. "I definitely liked doing the job, but there's no doubt about it, at that age you do make mistakes. Everything doesn't work out the way it does on paper, that's for sure."

Fate will always come up with its challenges and it wasn't long before Cannon Hall Farm was taking on a new role – as a refuge for his brother-in-law's animals. Areas of Laurie and Shirley's farm were built on top of a mine shaft and overnight holes began to start to appear in the farmyard. Their farm animals needed to be rehoused quickly while their land was made safe, and Roger and Rene had just the place. Suddenly, the farm was filled to the brim.

Laurie had always kept horses and, along with a new consignment of pigs, Roger took over the care of Silver, a fine semi-thorough-bred. Having been a keen rider as a young boy with his pony Rosie, Roger now had a new trusty steed to care for. So after a day's hard work farming out in the fields, he liked nothing better than a canter on Silver through the nearby woodland. "I could set her off at quite a pace and would ride her bareback through Deffer Wood. I was never one of those lads who was into motorbikes, so Silver was my version of a motorbike, I suppose – my bit of danger."

7

Love and marriage

E ven though he had to support his mother and run the farm, Roger was still able to find time to enjoy himself and, being sports mad, he kept on playing cricket and football. He joined the Cawthorne cricket club when he was 16 and even got into the final 14 at Barnsley Boys, training twice a week at the impressive Shaw Lane cricket ground.

He also became more involved with the Young Farmers' Club, which was not only great for his social life. Well known for being an opportunity to spark up romances, it also helped young people develop their farming skills. At the age of 16, and again the following year, Roger was chosen as a young cattle judge to represent Yorkshire at the annual Dairy Show held in London.

It was through Young Farmers that Roger met Cynthia Dickin, the woman who was later to become his wife. He had made friends with Cynthia's brother Ted at a Young Farmers' quiz night and he had invited Roger to come along to a Young Farmers' dance that was taking place in Halifax a few weeks later.

But come the actual night of the dance in January 1962, Roger very nearly cancelled. The weather was frightful: several inches of thick snow had settled and there was a bitter wind whistling outdoors. Suddenly, staying home with a nice cup of tea in front of the fire seemed quite tempting. But when you're a strong, healthy 19-year-old – and more importantly, single – a bit of bad weather isn't going to hold you back, even if it does involve a 52 mile round trip to Halifax.

Perhaps not surprisingly, there wasn't a huge turnout at the dance that night. Only a couple of dozen people had braved the elements to get to Spring Hall, so it wasn't exactly a rip-roaring atmosphere. The compere tried his best, though, playing records and coming up with games to liven up the night.

One cheeky competition offered up a prize for the first girl who could take off one of her stockings and bring it to the compere at the music desk. Despite the Arctic conditions outside, all the girls there took part – including Cynthia.

Like Roger, 17-year-old Cynthia had thought twice about going to the dance that night, but her brother Ted had convinced her it would be fun. She was in a bad mood because her driving test had been cancelled that day due to the weather. She had been looking forward to new independence, rather than having to ask for lifts all the time.

Instead of sitting and sulking at home, Cynthia reluctantly agreed to go along with Ted. She wasn't the twin-set-and-pearls type who dreamed of meeting Mr Right and getting swept off their feet at the dance hall. For Cynthia, it was just something to do in the middle of winter. But when it came to competitions – even one as daft as a stocking race – there was no match for Cynthia's determination. When the compere said, "Get set, go!" she had her stocking off and reached him quicker than you could say Barnsley chop.

Coming back from the stage with her winner's prize – "It really wasn't worth the effort," she recalls – Roger was grinning at her from ear to ear. "Can I help you put that stocking back on?" he asked her. Cynthia, never one to suffer fools, was not impressed. She replied with a scowl: "No, thank you."

Her frosty expression would have sent most boys packing, but Roger was undeterred. You never get a second chance to make a first impression and Roger wanted her to know that he wasn't just a cheeky so-and-so. He was going to show her that he was worth getting to know. Cynthia wasn't convinced, but she let him get on with it. "Well," she remembers, "there wasn't a great deal of choice that night. There were only about 25 of us who were daft enough to turn up."

The night continued with more dances and jolly competitions, and somehow Roger convinced Cynthia that it would be worth her while to meet him again. "I remember seeing my friend Pauline the next day," Cynthia says, "and saying, 'There's a silly fool coming all the way over from Barnsley to see me.'"

When news reached Cynthia's parents that she had met a potential boyfriend at the dance, they weren't, she recalls, exactly thrilled. Owning a sheep and poultry farm themselves, Olive and Ted Dicken were hoping their daughter wouldn't end up being a farmer's wife. They'd always aspired to a better future for her.

Cynthia's life may well have gone in a different direction had she not met Roger. When she left school at the age of 15, Cynthia went to college in Leeds and learned how to work a mechanical calculator known as a comptometer, before landing a clerical job at Dean Clough carpet factory in Halifax. Considering she had been quite a rebel at school, Cynthia was carving out a nice career for herself. "I don't think I was very clever and when I got a chance to re-take my

11-plus and go to a better school, I refused because I was having too much fun at Batty Road [Battinson Road School]," she remembers. "On the day I was meant to sit the exam I 'forgot' my glasses."

Sure enough, a few days after the dance, Roger went to Halifax to meet Cynthia. They had agreed to meet outside the General Post Office, which back then was everyone's favourite meeting place. Despite the less than auspicious start to their relationship, Cynthia began to warm to young Roger and was impressed that he had taken on the responsibilities of running Cannon Hall Farm at such a young age. Perhaps he wasn't such a Jack the lad after all? Roger, meanwhile, had fallen for Cynthia's sparkly brown eyes, her confidence and her dry sense of humour. A girl like her would certainly keep him on his toes.

After a much more successful date than their first encounter, from that moment on, Cynthia and Roger were an item. Roger would happily brave any weather to battle to the top of the thousand-foot hill to Norton Tower, where Cynthia lived. "I remember walking up there one day and the snow was so bad it was banking up the side of me," he says. "I had to plough my way through the drifts to get to her house. When I got to her house, the snow was above the front door and I had to tunnel my way in!"

But come hell, high water or snow drifts, Roger was determined to show Cynthia that he was the one for her. There would be trips to the cinema – "I always liked those Kirk Douglas macho movies and cowboy films," says Roger – or maybe a meal out. "We used to go for a Chinese and we'd always have the Far East special, which was served with a fried egg on top." And there were always lots of Young Farmers' events, such as Saturday-night dances – with or without stocking races.

Roger would meet Cynthia off the bus from Halifax every weekend and she would stay over at the farm, helping out with the various jobs that needed doing and riding on the back of Roger's tractor while he was turning the hay. "She wasn't so keen on cutting turnips when it was wet and cold in the middle of winter, though," Roger remembers. "I think that may have put her off me a little…"

Three and a half years of courting later, he finally proposed. "He didn't go down on one knee or owt soppy like that," Cynthia says. "We got engaged in The Spencer Arms in Cawthorne and the ring was from Fillans Jewellers in Huddersfield. I remember he bought me a second-hand ring so that he could save on purchase tax."

"Aye, that meant it were better value!" Roger laughs. "And I made sure I knew she'd say yes before I bought it. I wasn't going to risk forking out for nothing!"

The wedding itself was very nearly a disaster. Just a few days before Cynthia and Roger were due to tie the knot, they discovered that the catering company had run off with their money. Everything had been arranged, from the venue, the table flowers and napkins, to the fancy three-course meal for 100 people. But when the couple tried to finalise the details and the phone kept ringing out, they knew something was wrong. They then heard through the grapevine that the wedding organisers had done a moonlight flit.

"We had nowhere for the reception, as we couldn't find anywhere with such short notice," says Cynthia. "We ended up having to book a place that was miles away from the church and we had to have a buffet brought in, rather than a sit-down meal."

Things weren't going to plan. "I'd wanted a really plain frock, nothing showy or fancy," says Cynthia. In the end, she wore a beautiful dress that was the perfect combination of simple elegance, trimmed in scalloped lace with long sleeves and a scooped neckline. It was topped off with a pretty headpiece and a short lace-embroidered veil. "Don't get me wrong," Cynthia laughs. "I was really happy to be getting married, but I didn't like all the fuss. I just went on the bus to get my hair done – there was none of this daft stuff of getting a make-up artist in to do your face and fiddle with your hair. You just sort of got on with it. I was the last to start getting ready and the first to finish."

On Saturday 2 October 1965, Roger Nicholson and Cynthia Dickin married at Saint John's Church in Warley, Halifax. Photos from the day show the two of them looking very much in love as they pose for photographs with their relatives and Roger's best man, Alan Nichols. Roger looks very handsome in his charcoal grey suit, while a radiant Cynthia in her stylish horn-rimmed glasses mirrors his romantic smile.

Roger and Cynthia
are all smiles on their
wedding day in 1965

8

The family grows

No matter how much in love the newlyweds were, having had to fork out for two receptions meant a honeymoon was out of the question. And it wasn't just a financial decision. When you have a busy farm to run, you can't just drop everything and head off on holiday.

The couple had opted to marry in October as it was the quietest time in the farming calendar, the idea being that they could enjoy some time together while Cynthia settled in. That year, however, the weather was against them. "We'd got no crops in. Nothing," says Roger. "So I went straight back to work the day after we married and the corn was black. When the combine went in there, there were clouds of dust and muck. It was awful. Then I got the flu."

"I'd barely taken my wedding dress off and I was straight out there in the fields bagging corn from the combine," says Cynthia. "I thought if this is what it's like being a farmer's wife, I may have made a big mistake!"

Joking aside, and despite her parents' initial misgivings that it wasn't a good idea to marry a farmer, Cynthia never had any doubts. She knew that being a farmer's wife was going to be hard work, but it never bothered her in the slightest.

Now there were two Mrs Nicholsons at Cannon Hall Farm, and although Roger was always the apple of Rene's eye, the two strong women rubbed along nicely together. They were both no-nonsense people and the fact that Cynthia had a farming background meant that she wasn't fazed by any of the more challenging sides of rural life, such as occasionally losing animals during lambing or having to send animals to slaughter.

With Rene living in the main farmhouse, Roger and Cynthia moved into Tower Cottage to start their married life. Thanks to Cynthia's old colleagues at Dean Clough's, the property had been kitted out with the best carpets that money could buy and they had good quality furniture. But that's where the luxuries ended. With no cooker downstairs, the young couple would go to Rene's house to cook their meals, then go back to Tower Cottage every night.

It was an unusual set-up, but it seemed to work. Well, most of the time. "One night we went back to Tower Cottage and left the door open at the farmhouse by mistake," Roger remembers. "A cow walked straight into the farmhouse and put his feet on Mum's bed. I think it's safe to say she wasn't impressed."

Rene had always looked forward to the day that she would have grandchildren, and she didn't have long to wait. The following July the newest member of the Nicholson family was scheduled to join the family, only the baby was a bit reluctant to make his debut appearance... "I was 10 days overdue when the doctor said, 'Enough's enough, we're taking you in'," Cynthia says. "I couldn't have been more relieved."

Cynthia gave birth to a whopping baby boy weighing nine pounds and one ounce, who she and Roger named Richard Charles. Because the baby was induced, it was a quick, straightforward labour, but Roger didn't stick around to cut the cord. "I'd seen enough birthing in my life at the farm, so no thank you, I didn't want to be there. I still think it's ridiculous that you'd want to be around to see it now!"

Instead, Roger went to wet the baby's head with his sister Shirley and visited later on when all the messy stuff was over. "In those days you stayed in hospital for 10 days after the birth," Cynthia continues. "Which was just as well – when they examined me, they discovered they'd left part of the afterbirth behind, so they had to cut me open again."

Roger didn't like the fact that all the new mothers were kept in one ward while all their babies were in a side room. It seemed to him that the only bonding took place when the babies were being fed. "As a bit of joke, I said to them, 'I look after my lambs better than you look after these babies!' But, oh dear me, they didn't take kindly to that at all. They hauled me into the sister's office and asked me if I was making an official complaint. I had to explain it were only a bit of frivolity!"

Back on the farm, things took a while to settle down, as baby Richard suffered badly from colic and didn't stop kicking and screaming for several weeks. Being a July baby, at least Cynthia could push his pram around the village in the sunshine while he wailed his lungs off, and she was glad of the extra help from her mum Olive and mother-in-law Rene. "Even though it all happened quite quickly, I wasn't thrown by the noise and the nappies," says Roger. "I was very proud to be a father and I'd always planned to be very hands-on if I had any children."

When Roger had been tiny, his dad Charlie used to take him along to the market when he was selling cattle, so Roger kept up the tradition with baby Richard. He remembers a day when he was selling some lambs and the auctioneer said, "Come on, everyone, he's got a family to feed here! Let's see those bids!" Richard was only about four months old at the time, but he was as good as gold that day. Perhaps he would grow up to be a farmer too?

Less than two years later, in April 1968, another son, Robert Edward, came along, followed by David William two years after that in April 1970. "I wouldn't have minded having a girl somewhere along the line," says Cynthia, "but I was just relieved to have three healthy boys. When David was born, the midwife said to me, 'Oh, I am sorry'. I thought something terrible had happened and started panicking. Then the midwife said, 'No, there's nothing wrong. It's just that you have another son.' As if that was going to bother me."

At that point, Roger and Cynthia decided that three children was plenty. "I hadn't a lot of money coming in at the time, so any more would have just been irresponsible," Roger says. "As it was we had three children under the age of four, so life was never quiet." Luckily Cynthia and Roger had lots of support from Cynthia's parents, who would regularly come over to Cannon Hall Farm to help out with the demanding brood.

As time went on, having been a very noisy baby, Richard started growing into a placid little boy, happy to quietly play with his toys. "He was a real pond dipper," remembers Roger. "He'd like to grovel around in the grates by the side of Tower Cottage and see what he could find. Then he'd collect a bucket full of frogs, newts and toads and watch them for hours before putting them back."

"I remember being quite disturbed when a huge beetle crawled up my arm," Richard recalls. "But then I started getting interested in insects and I'd dig into the earth and turn stones over to see what I could find." It was to be the start of Richard's life-long fascination with unusual mini-beasts.

Meanwhile, Robert, their cheeky blonde-haired second child, was a little firecracker. From the moment the tiny adventurer could walk, there was no stopping him. "We had to get reins for him," says Cynthia, "we were that worried that we wouldn't be able to catch him! Robert's favourite thing to do when he was a tot was to let all of the hens run out into the farmyard, then chase around after all of them and put them back in the shed. Then, once they were all in, he'd let them all out again. He'd do it over and over again, and the poor Bantams would be running their legs off."

By all accounts, David, the youngest, was a very good baby – "Mummy's pet rat, they used to call him," says Cynthia. Even so, the three boys took some watching when they got together. "It was a cracking place to grow up," says Robert. "There was always so much to do and as soon as we could, we'd be off building dens and digging tunnels, catching hens and playing with the farm animals."

As a special treat, one day Grandma Olive took the three boys to see the new *Robin Hood* Disney film at the cinema, buying them all bows and arrows as souvenirs. "I think she realised her mistake quite quickly," says David. "We went back to her house after the film and started using her prized china ornaments for shooting practise. It was so funny running around her lounge firing at her pot ducks. Then there was a lot of shouting and our bows and arrows were confiscated. Funnily enough we never saw them again."

Living next to Cannon Hall, with its miles of country parkland, streams and wooded areas, meant there were plenty of places to explore and have adventures, and the three Nicholson boys made the most of every second. And as their confidence grew and they watched their dad while he worked, they learned more and more about how to look after animals. "I can't think when the very first time would have been that I helped deliver a calf or a lamb – it's just something we always did," says Robert. "Dad would get the new arrival into the correct position and we'd help guide the legs out and wipe the new animal down with straw. And when we were much younger kids, we always had broody hens and we'd slip eggs underneath them and wait for the chicks to hatch."

Pigeon racing in Yorkshire also presented the boys with opportunities to get hands-on. "We used to get quite a lot of strays over our way," says Robert. "So we'd catch them, feed them up and rehabilitate them, then let them fly back home again. It just sort of came naturally to us. We had one that I particularly remember called Kez [named after the kestrel in the 1970s film] and we kept that one for a long time. If we could make a pet of something, we would."

It was no wonder that the children had such an affinity with animals. Their mum Cynthia had always loved them, too. At the age of 12, she bought herself a calf and used to take him for walks – which was all well and good until Billy the calf grew into Billy the bull... "My brother Ted and I always had loads of farm animal pets, but they always seemed to be getting out of their pens. I remember having to miss school on a number of occasions to go and find them!"

From a very early age, Robert showed signs of his entrepreneurial spirit when he happened upon a way to make some money from their pet hens – or possibly, he admits now, it was more a juvenile form of extortion. "We'd stand at the front of Cannon Hall and – because he was the smallest and looked the cutest of the three of us – I'd get David to hold a baby chick and then charge people 2p a stroke. We actually made a fair bit – apart from the people who just tutted at us and didn't want to know!"

"He'd then collect all the takings," adds David, "saying that way we'd have a nice pile of money to spend on sweets. Then he'd change his mind and take most of it for himself and only give me a couple of pennies!"

Another of Robert's money-making ruses was to collect together any old eggs that he could find around the dusty nooks and crannies of the barn – not the fresh ones, they would already have been collected for breakfast – which he would then box up and try and sell to visitors in the country park. He'd make sure the buyer wasn't anyone he knew from the village, or anyone who'd be able to track him down and complain, because some of the eggs were likely to be several weeks old.

And then there was the time when the boys' guinea pig Guinness turned out to be a pregnant female and gave birth to a litter of five pups. Robert convinced the pet shop owner to buy each one back from him for £2. Robert showed that when there was money to be made, he could sell coals to Newcastle.

Together with their network of good pals from Cawthorne village, the boys would often go off fishing together with their dad. They would be off first thing in the morning with their ham sandwiches, nets, and little yellow bucket and follow one of the streams that would thread from the river into Deffer Wood. There they'd try and catch sticklebacks and bring the fish home to put in their garden pond. When they got a bit older, they were allowed to go off by themselves.

But not all of their adventures went to plan. One morning, Robert, David and their friends went off to enjoy one of their more dangerous pursuits – walking across the slippery stones in the shallows of Middle Lake's waterfalls. The two boys hadn't yet learned to swim and would have been wearing armbands had they been at a swimming pool. Then again, they hadn't actually been planning on a dip…

They were all nearly at the other side when David, who was just seven at the time, lost his footing and sunk like a stone. "I remember being fine one moment and then being swallowed up by the stream," he says. "The shock of the cold and the roaring of the water in my ears was terrifying and everything seemed to go into slow motion. Some kind of survival instinct must have kicked in though and I came up for air and somehow managed to doggy-paddle to the side."

Everything happened so fast. One minute, Robert was chatting away to his friends and having the time of his life, the next he was panicking and screaming as his little brother was evidently drowning in front of him. With no time to waste, the gang reached out to David and yanked him towards the bank, heaving him out of the water and on to dry land. Little David was in a bad way, but what should they do? They would be in for a hiding if any of their parents found out they had been walking across the waterfall.

Robert came up with a plan. While the shivering, dripping-wet David hid under a stone bridge in an area of the country park called Fairyland, Robert hot-footed back home, snuck into the house and found an almost identical set of clothes to the one David had dressed in that morning. Then he ran back and helped his little brother change out of his soggy outfit. They figured that if they could hide

52

the wet clothes somewhere, then sneak them back home when they were dry, their mum and dad would never find out.

That teatime, Cynthia wondered why her boys were so quiet. They were usually full of stories about their adventures and telling tales. Her sixth sense told her something was going on as her three little chatterboxes just sat at the table and ate in silence.

An hour or so later she found the sodden clothes that Robert and David had forgotten to hide.

"We would have got away with it," says Robert. "But Richard couldn't help himself and he told Mum what had happened. In hindsight, I know it was just that Richard hated keeping anything from her - he could never be deceptive, but at the time I couldn't believe he'd snitch on us…"

But instead of getting a hiding – though voices were certainly raised and the boys were forbidden from ever going near the lake again – Cynthia and Roger decided that the boys needed to be taught how to swim properly. The following weekend they were frogmarched off to Barnsley for swimming lessons. David's near miss had been a lucky escape. "Dad got a great deal on the three of us having joint lessons," says Richard. "But even though I was 11 at the time and already had hairy legs, I still had to go with them into the tiny kids pool. It was humiliating, but at least it meant I quickly learned how to swim!"

The three brothers with their fishing gear

9

Three boisterous boys

The boys were always surrounded by animals at home. Along with the predictable Old MacDonald varieties like hens, pigs, sheep and cows, they also persuaded their parents to let them have exotic pets such as garter snakes and curly-tailed lizards.

The boys adored their reptiles and were never in any hurry to put them back in their tanks at the end of the day. "We were always losing them," says David. "Mum wasn't best pleased if we had to ask for her help to look for them. Once we found Sid the snake in the back of a wardrobe and another time I woke up in the morning to find him coiled up on the top of my head, as they always head for the hottest place they can find. Once he went missing for a couple of weeks and did the same thing to my grandma - curling up on the top of her head while she was asleep. When she woke up to find him there she absolutely screamed the house down."

The three boys epitomised the word boisterous. One evening, when Cynthia and Roger went out to play badminton and Grandma Rene was babysitting, the boys were curiously quiet. "We said, 'Night, night, Grandma' and snuck out to the turnip chopping barn," says Richard. "There was a hole in the barn door and we'd decided it would make a great launch pad for a firework. It was a few days before Guy Fawkes Night and we'd pooled our pocket money to buy a rocket. We wedged it in and lit the fuse…

"Well, it was a lot more powerful than we'd thought and it went 'Whooomph!' straight across the farmyard. It gave Gran the shock of her life. She called to us saying, 'There's some bad boys in the farmyard letting off fireworks, I'm going to call the police!' And we had to beg her not to. She probably had no doubt at all that it was us, but just wanted to put the fear of God into us. And it worked. We wondered if she was going to tell Mum and Dad about it, but she never did. Maybe she was keeping it in her back pocket for another time we might need teaching a lesson? Of course, we never should have done it because we could have burned the farm down…"

"One of our favourite things to do was have rotten egg fights," says David. "We'd find the eggs in various areas of the barns where the hens didn't normally lay and collect them together, then shake each one to find out if it was bad. If it sounded watery and sloppy inside we knew they were ancient and they'd make great mini stink grenades."

"We'd invite our friends over and have two teams," continues Robert; "then we'd divide up the eggs and pelt them at each other. They'd explode with a terrible plop and the smell was horrendous. Unfortunately one day a sales rep was visiting Dad and one of the eggs went through the open window of his car and smashed on the dashboard. I grabbed my friend's jumper and mopped it up, but I don't think it was very effective. Then we all hid and watched from afar as the rep got back into his car and drove off. We got away with it though…"

If they weren't causing mischief, the three brothers enjoyed practising their sporting skills. Like his dad, Robert was cricket mad and would get his patient big brother Richard to bowl for hours and hours while he perfected his batting technique. To make things a bit more interesting, the brothers devised an extra scoring system: one point if you hit the roof of a building but still managed to catch the ball; out if you hit the milk churn wickets or broke a window.

Richard was less keen on perfecting his batting. "I was always much happier playing indoors and drawing to be honest," he says. "I'd invite my pals over, then I'd get bored with them and leave them outside playing football and cricket with Robert and David."

Whatever they were doing, from playing indoors, going to cubs and scouts, riding their bikes to buy sweets in Cawthorne village, milking the family cow – called Pantomime Cow because of its droopy udders – or getting muddy with the pigs, the three boys lived life to its fullest.

With the boys growing up at a rate of knots, Roger worked as hard as he could to support his family. He tried his hand at every possible kind of farming he could think of, striving to make a good living for the family. "I'd never had any yearnings for a fast car or fancy holidays," he says. "For me, the main challenge was finding a system to make enough money for us to live. I tried everything under the sun, really."

Early on, Roger had hoped to become a dairy farmer, but then he discovered how much money he would have to spend on a new dairy system and larger herd in order to make it work. "The milk we obtained from our small herd went straight to the Milk Marketing Board – nobody had artisan cheeses or owt like that in those days – and we'd get about £700 a year for it," he explains. "True, £700 bought a lot more back then than it would do now, but £700 is not much for all that work – milking twice a day for 365 days a year. We didn't seem to be getting anywhere. We needed to be milking 70 or 80 cows at least to make a proper living from dairy farming. At that time, we only had 10 cows, so we had to scrub that idea."

Next he tried pig farming, rearing the traditional Large White Cross Landrace varieties, the standard choice for pork at the time. "I just couldn't find the secret to making it work," says Roger. "The buildings that we had at the farm weren't the right sizes for doing it in a big way – we would have needed to rebuild and we just didn't have the money to do it. I remember going to market one day with my pigs and being really upset about the pittance I could raise for them. I remember saying, 'Surely they must be worth more than that for all the work I've put in?' I was proper down about it because other farmers around me seemed to be able to make a living from it, but I just couldn't."

As is the case for many other farmers, there were times when they really struggled to make ends meet. One afternoon, Cynthia and the boys were watching television, when there was a knock at the door. Two burly men were on the doorstep saying they'd come to repossess the TV set as the rental on it hadn't been paid. "I said, 'Couldn't you at least wait until the end of Scooby Doo?' Cynthia says. "But they weren't having it. They just unplugged it and took it away."

"I remember we were all laughing about it," says Richard, "until we realised we'd no telly to watch…"

Luckily, being a family of sport lovers, there was always football to be played – often in the house with doorways for goalposts. Board games were also a regular feature, with the boys showing at an early age that they had a very competitive streak. "Sunday night was always my favourite," says Cynthia. "We'd play Yahtzee and Connect Four and cook pizzas with 'toppings of your choice'. It wasn't easy, though, as the boys all liked different things, but we'd get there eventually."

Happily, when Christmastime came around, there was one particular revenue that Roger and Cynthia could always depend upon.

"Turkeys were probably our most reliable source of income – one year we sold 700 of them," says Cynthia. "Roger and I used to have a good little production line going. Once the turkeys were killed, we'd pluck them straight away and hang them for a few days – that way they taste better – and then gut and truss them so they were ready for customers."

"We had no fridges when we started out," says Roger, "so we were always hoping for cold weather to be able to store them safely and to get them ready in time to deliver them all by Christmas Eve. I used to get five bob a pound for them."

It would be pretty much non-stop for the two weeks leading up to Christmas Eve, often with extra family and friends getting pulled in to help get the orders out. "It was so busy that we knew if I didn't get the Christmas tree up by the 10th December it wouldn't happen," says Cynthia. "We'd be working so hard getting the turkeys ready that after a while you couldn't feel your fingers."

Selling to local butchers and around a hundred private customers around Cawthorne village meant at least Roger could afford to get the boys a present for Christmas. With the last turkey delivered on Christmas Eve, and maybe a quick drink to toast everyone for their efforts, Roger would then battle with the other last-minute shoppers in Barnsley to buy the boys a big joint present. One time it was Subbuteo, another time a snooker table. "The important thing was that it would be something we could all play with together," he says.

From grainy black and white to glorious technicolour, the Nicholsons' family photos show the boys from tots and tweens to teens in their pyjamas on Christmas morning, tearing into their Christmas presents and posing at parties with various members of the family, including Grandma Rene and great aunty Flo, who always entertained everyone with the funniest dirty jokes.

Lifelong friends Rosemary and Nigel are in many photos, too. They often joined in the turkey production line and would come around on Christmas Eve, staying past midnight, when it was technically Christmas Day, so they could watch the children open some of their presents. "Then they'd disappear home," says Cynthia, "and we'd be left with these three hyperactive boys who were too excited to go back to bed and just wanted to stay up playing!"

"We used to love Christmas Eve," says David. "We always had a party and Rosemary and Nigel would play three card brag and pontoon with us for one-penny and two-penny bets. Then early the next day we would leave all our presents behind, get in the car and travel to Halifax to see Grandma Olive and Granddad Ted – Mum's parents. I remember we'd always have a starter of Yorkshire pudding with onion gravy, even when we were about to have a massive meal of turkey and all the trimmings."

Rosemary and Nigel feature a great deal as the story of Cannon Hall Farm unfolds. Nigel helped Roger on the farm and Rosemary often went on holiday with Cynthia and the boys. Because of the demands of the farm, it wasn't possible for Roger and Cynthia to go on holiday with the children at the same time. "We tried it once and it was a disaster," says Cynthia. "We borrowed a car and drove all the way to Paignton in Devon while our friend Brian looked after the farm while we were away. We'd only been gone a day when he called us to say a pipe had burst and we had to come straight home. That was the end of that holiday."

The following year, and for many summers afterwards, the boys either holidayed with Cynthia and Rosemary, or went away with their dad. "Happy times!" Rosemary smiles. "We'd stay in a B&B in Filey or Scarborough and have picnics, go to pitch and putt, or fishing in little rock pools. Cheap and cheerful holidays with the five of us staying in one big family room. We all loved it."

"I remember every single one of our holidays – mainly because there weren't that many of them!" says Roger. "When I used to take the boys away with me, I'd always say to them, 'We can stay in this place which is a bit posher, or we can stay in this cheaper place and we'll have more money for slot machines and crazy golf', and the boys would always choose the cheaper option. I remember some of the hotels were terrible places, though. One night at one of the places we were staying at, the fire alarm went off and Robert tried to convince the old ladies staying there that David had caused the alarm to go off by sweating so much in the nylon sheets!"

Life with the three boys was always colourful. "Cynthia once asked me to babysit for the boys as she wanted to visit Roger in hospital after he'd had a minor op," Rosemary says. "I remember saying, 'How about I visit Roger and you stay home with the boys! They were quite a force of nature."

"We had running battles with the head gardener at Cannon Hall," Richard recalls. "He was always chasing us off his perfectly manicured lawn." But even Roger had a run in with the grumpy groundsman over a stray cricket ball, so perhaps the boys weren't the only naughty ones.

Meanwhile, Grandma Rene had got the travelling bug. Having led a very quiet life, in the most literal sense, getting hearing aids in her late fifties opened up a whole new world to her. She joined the women's branch of the Rotary Club, the Inner Wheel, and made plenty of new friends, going off on as many holidays as she could. "At 70 years old, she took herself off to New Zealand," says Roger, "and there were trips to America and Ibiza and a Mediterranean Cruise. She certainly made up for lost time."

Rene's new-found jet-set lifestyle wound down as she got into her eighties. Roger recalls that he began to have concerns about her driving. "I could see that something wasn't quite right," he says. "The car seemed to know its way home better than my mother did."

Concerned that Rene might be putting herself in danger, Roger took over the driving service for her. Gradually, he began to realise something was amiss and although she was never technically diagnosed with Alzheimer's Disease, she showed all the symptoms. "She was around 85 when the Alzheimer's started," Roger explains. "She was gradually becoming more forgetful, then she had an aggressive period, then she was placid again." After a while she needed to have 24-hour care, so the family would stay with her all day and carers would come in for the night shifts.

"For most of the time she would just sit there regally, getting on with her life, and she was really quite comfortable until the final stage of her life." Roger recalls. "We were fortunate that she always knew who we all were and she loved spending time with her grandchildren and great grandchildren. It could have been a lot worse."

Sadly, in November 1995, Rene passed away, aged 91. "Cynthia and I had gone in to turn her in her bed, but she didn't want to be fussed over. Later that evening she died. I held her hand, but when she didn't grip it back I knew she had passed away."

"I suppose when someone gets to 91 and has been through a good retirement, then it's not as sad as other circumstances," says Roger. "Looking back at her life and our relationship I was so lucky to have a mum like her because she always backed all of my decisions about the farm. She never once told me that I was daft for trying to make new ways to make a living and we certainly had plenty of happy times."

One of the family's favourite times of the year was the bi-annual Cawthorne summer carnival. All the villagers and families from surrounding areas would get involved, making fantastic floats and costumes based on different themes each year. "One time we decided to be The Royal Mail and Cynthia and I made ourselves into post boxes with cardboard frames hanging from our shoulders," says Rosemary. "Our float had fantastic hand-painted horses pulling a traditional mail coach, complete with wooden cartwheels and a sign that said London to Cawthorne. We won first prize that year – and the Mayor's Parade!"

Another year saw the two friends transforming themselves into Bill and Ben the Flowerpot men, complete with plastic flowerpot legs, rosy cheeks and a smiling sunflower. "The trouble was we couldn't bend our knees," laughs Rosemary, "so we had to be lowered into a sitting position." Then there was the year that Cynthia and

Rosemary made themselves into a Spanish onion and a Seville orange. They were highly delighted with their papier-mâché creations – until the time came for them to reveal their costumes and they discovered they were too big to get out of the doorway.

Families would spend months beforehand building and planning the floats, and everyone got involved. It seemed many people embraced the chance to dress up and anyone with a sewing machine was kept busy running up costumes of everything from Maid Marian to Boy George. Archive amateur films of the Carnival during the Seventies and Eighties show dozens of children kitted out as characters from *Chitty Chitty Bang Bang* and Ali Baba's cave attendants, as well as nursery-rhyme favourites, superheroes, cowboys and gangsters. "One year our float was called 'Oh I do like to be beside the seaside'," says Richard. "Mum knitted Victorian swimming costumes for me and Robert and we had little straw boaters to match. I still have the embarrassing photographic evidence!"

Being England, the only thing people could rely on was for the weather being unreliable, but nevertheless everyone joined in the fun. Roger dressed up as a very cute baby one year, complete with dummy, nappy and pretty pink baby-grow. And another year he couldn't resist dressing up as Tommy Steele's Little White Bull. After all, there was a family connection...

But like so much of village life, the bi-annual carnival fizzled out. What started out as a village jamboree that everyone got involved with, got bigger and more professional as time went on. Gone were the days of kids' wonky signs and cobbling together costumes from whatever you could find at home. It was turning into a much more corporate affair. Bigger businesses started getting involved and the floats looked more like stage sets. One year a stunning Treasure

Island galleon was so huge that the telephone wires had to be lifted in order to get it through the village. It all became too competitive and what had originally just been a bit of fun for the family to join in became more of a chore.

When the Nicholson boys went to the school at Cawthorne, most of Cynthia and Roger's social life was based around the village, much of it with Rosemary and Nigel. The four were members of the tennis club and had a long badminton career together, often coming face to face to battle it out in the finals. And there were dances, parties and the annual harvest festival with a pie and peas supper.

It was another excuse to get dressed up and let your hair down. "They were great times," says Roger "and we'd provide all the entertainment ourselves. I did all sorts of things, like doing a *Come Dancing* ballroom routine with an inflatable doll tied to my shoes and singing *Are You Lonesome Tonight* dressed as Elvis." One time Richard drew a schooner on Roger's tummy for the finale of *South Pacific's There is Nothing Like a Dame*, but the funniest time of all was when Roger would get together for duets with his friend Ron. "One time we did the *Indian Love Call*," says Roger. "He was at one end of the village hall dressed as a Mountie and I was at the other dressed as a native American princess singing 'When I'm calling you - hoo he hoo...' I'm sure it was a bit pathetic, but everyone used to laugh like a drain!"

"In the days of the dances at the village hall, Cynthia and Nigel would prop up the bar having a drink, and Roger and I would be on the dance floor," says Rosemary. "Roger and I had this jiving routine to *Rock Around the Clock*. He'd throw me around his hip three times and the fourth time we'd always end up on a pile on the floor. Then Roger would go outside and be sick!"

"She took a lot of spinning," adds Roger.

As their friendship grew, Cynthia and Rosemary became known as Flo and Con. Rosemary explains why. "When I left school, I did a cookery course and I got the nickname Con after the cook Constance Spry. And Cynthia got her nickname for her nursing skills."

In 1984, during the miners' strike, the foursome were out collecting timber for pensioners in the village when a branch fell and injured Rosemary's husband Nigel. Cynthia came to the rescue with her medical know-how and the Florence Nightingale nickname stuck. Everyone took to their new nicknames – except for Rene, who sternly used to say, "I don't know why you two call each other Flo and Con. You've got two perfectly good names – use them!"

In 1977, the Queen marked her Silver Jubilee year with a national and international tour, travelling over 56,000 miles around the globe. During their tour of South Yorkshire, one of the places they chose to visit was Cannon Hall and crowds lined up to wave their Union Jacks, dressed in red, white and blue.

As the Nicholsons' house was just behind the Hall, it was a nice stroll over to take up their positions to see the Queen and Prince Philip on their walk of the lawns. David remembers being seven years old and getting pole position right at the front. "There was an old veteran next to me and after the Queen went past he turned to me and said, 'You stood there very well, young man!' And he gave me five pence!"

There was more excitement to come when the Queen's security

team parked the Land Rover that carried the Queen's Lady in Waiting in the courtyard outside the Nicholsons' house. David continues: "Grandad Ted [Cynthia's father] asked if the security team would like to have a cup of tea while they were waiting for Her Majesty, and as a thank you, they gave us a drive around the courtyard."

Robert, meanwhile, was there on a school trip. "I know it sounds ridiculous now, but I somehow managed to get lost!" he laughs "There were thousands of people there and me and my friend Darren got separated from the school group and took ourselves to the lost children's tent. The next day we got a right telling off from the headmaster in assembly. He said, 'Two boys really let the school down – and one of the culprits actually lives there!' I thought I'd been a very good boy by following protocol, but I obviously got it wrong!"

10

Reaching adulthood

As each of the boys finished at junior school in Cawthorne, they followed in their Grandfather Charlie's footsteps and went to Penistone Grammar Secondary School in Barnsley. The school motto at the time was *Disce aut discede*, meaning Learn or Leave, so the boys knew their days of pranks, pets and farm adventures would have to be put to one side – during class time at any rate. It was time to knuckle down and get some qualifications. But then again, boys will be boys…

So what did the boys make of their school years? "For the years one to five, I pretty much hated school," says Richard. "I was saved by lunchtime table tennis and the art department." Nevertheless, the oldest of the Nicholson brothers did well in his O levels and went on to study geography, biology and art in sixth form. "I found I could never focus on geography and biology, though, so I dropped them and just concentrated on my art. I remember it was quite an honour for the headmaster to have one of the pupil's artwork in his office and for a long time he had a picture of a bull that I had drawn hanging on his wall."

Living on a farm certainly had its uses. "Dad knew the local abattoir owner," says Richard, "so my science teacher would ask me

69

to bring in hearts and eyeballs for the pupils to dissect." Luckily, the art teacher didn't insist on a real bull as a life model.

Meanwhile, for Robert, senior school was all about sport. "I started off in all the top sets when I got there and I remember being really good at sciences, maths and English, but by the time I left I'd decided I'd rather just be on the farm and playing sports." Getting his colours in the football team, table tennis team and being in the winning team for the Barnsley schoolboy cricket cup final, meant that Robert may not have scraped many O levels, but he has good memories of his time at Penistone Grammar – even if his school days did end rather abruptly.

"I managed to get suspended on my last afternoon," he says. "We went to the pub at lunchtime for a few jars of cider and Mr Morrison, our PE teacher, literally sniffed out our crime before our very last lesson. He picked us off one by one, saying, 'You, you and you – you're all suspended!' I was gutted because I wanted to play football for the very last time with my classmates, but there was no reasoning with him. So I disappeared in ignominy, never to be seen again."

And as for David? "I hated school and wasn't academic in the least," he says. "I was very quiet and kept myself to myself until about third year, then I got challenged to a fight."

Picture the scene: a bunch of lads waiting for the school bus and everyone larking around as usual, then – pow! – David gets punched in the face. "I was at home poorly after that and I had about a week off school," says David. "And there was a rumour going around that it was because I'd been punched, but I'd been secretly practising my boxing combos in the mirror at home." Sure enough, when he got back to school, the fight was on. "We tussled for a few minutes, then

I went in with the punches and the other lad hit back with some karate kicks. Meanwhile, a teacher was frantically trying to unlock the door to get in and break us up." Finally, it was too much for David's opponent, who said he'd had enough. The gang scarpered and Rocky Nicholson hung up his gloves. "I never started the fight," David says. "I was never the aggressor, I just put it to bed."

The fun continued at home. "I remember one day Rob came back from Barnsley with a brew-your-own beer kit," says David. "We found a secret place in one of the farm buildings where we could concoct it without Mum or Dad finding out, but it was too cold for the yeast to grow. So we fixed up one of the heaters from the fish tank and set it at the right temperature, then added loads of sachets of sugar to help the flavour. Three weeks later, when we went to check it, there was a massive crust of mould on the top of it – it looked absolutely disgusting. Even so, Rob and Richard invited all their mates round one night to try it. It tasted like vomit, but that didn't put them off. And they were all off school for the next two weeks with throat infections. I think it may have been the beer..."

When they weren't up to mischief, the boys were getting a full education about farming back at home, helping out their dad whenever they could – especially during lambing time in spring.

David can vividly remember the first time he had to do a delivery on his own. "It was triplets! Rob, Richard and Dad had gone to see Barnsley FC play at home and before he left, Dad said, 'Keep an eye on this ewe for me'. He'd put her in a pen by herself and I went out to check on her and could see she was struggling a bit.

"I'd watched Dad lambing so many times before, so I knew what to do, but this was the first time I was completely on my own. As each of the three lambs was born, I cleared their airways and sprayed the navels with iodine so they wouldn't get infected, then stayed with them to make sure they were okay. I wasn't nervous at all. I was just so happy that I'd done everything right; Dad was so pleased with me when he got home. And Barnsley beat Sheffield Wednesday 3-2, so everyone was happy!"

While Robert and David were always keen to be at the business end of a new birth at the farm, Richard was more interested in staying at home and drawing. Roger and Cynthia weren't at all surprised when he told them he wanted to study at art college after school, rather than following a career in farming. "I'm just not made the same way as my farming-mad brothers, I suppose," Richard says. "But when I lived at the farm, I still mucked out the pigs and did all the dirty jobs like everyone else. I just didn't have the calling the other two had."

After school, Richard moved to Hull to study graphics, specialising in photography, which was to prove invaluable in Cannon Hall Farm's future. Robert opted for Askham Bryan college near York, the largest farming college in England. He then did a Youth Training Scheme back at Cannon Hall Farm, helping to reclaim the land that had been used for open cast mining.

When it was time for David to decide on further education, he applied for Bishop Burton College in Beverly in the East Riding of Yorkshire to study for his National Certificate in Agriculture. "I knew that Askham Bryan, where Robert went, had a great reputation," says David. "But he'd warned me there weren't many girls there, whereas Bishop Burton had secretarial courses, floristry and equine studies. Yes, I know women do go to farming college, but I knew there'd be lots more at Bishop Burton!

"I also considered joining the army at one point, just before the

Falklands War, but the reality of putting my life on the line kicked in, so that didn't last long! I then considered some kind of forestry work, or rearing fancy chickens, but the truth was, I just wanted to be at home on the farm."

Roger, meanwhile, was happy for his sons to pursue whatever career they chose. "I certainly never pushed any of them to be farmers," he says. "Some people do that with their children, but I don't think it's fair and Cynthia and I always encouraged the boys to make their own decisions.'

Robert was the first of Roger and Cynthia's sons to marry. "Julie and I met in 1987 and got married a year and five months later," he says. "It was a big decision as we were still so young – I was only 20 at the time – but it just seemed like the right thing to do."

David was next. He met a hairdresser called Anita in a pub in Barnsley not long after he left college and they married in Jamaica in 1994. "It was perfect," says David. "I even managed to go water skiing the morning we tied the knot."

A few years later, Richard met his wife Maxine and the two married in 2003. Three years later, they moved to the family's other farm in the Gunthwaite Valley.

Maxine, who spent most of her life in nursing, was diagnosed with cancer in 2007 and sadly passed away two years later. Richard later met his partner Clare Weatherall, a manager at a travel company. "Clare has been a great support to me, and particularly to my son Marshall, through some really challenging times," he says.

With the boys getting older and able to look after the farm,

Cynthia and Roger could finally catch up on all those holidays they had never taken together. "The last time I'd been abroad was when I was 14 and I'd gone to Italy with my sister Olive, her husband Vinnie and my little nephew Steven," says Roger. "I had to get to 50 before I got to go anywhere else! I remember going to renew my passport and there was this teenage lad looking back at me."

The days of nylon sheets and crazy golf were long behind them when Roger returned to Italy for his first holiday abroad since that childhood trip. This time around they stayed at one of the world's most luxurious hotels, the Cipriani in Venice. "We were there with our friends Ron and Margaret and Rosemary and Nigel, and I couldn't believe how expensive it was," says Roger. "Rosemary had gone ahead and booked it and I couldn't enjoy it because we were so skint at the time. The next time we went away I was given an ultimatum – that I had to enjoy myself!"

11

A new direction

There were many times during Roger's early years at Cannon Hall Farm when he wondered if he would be able to keep the farm afloat. No matter what he tried – whether it was concentrating on one income source, such as making a living from dairy farming or breeding cattle, or trying a mixture of arable and livestock – the figures just wouldn't add up. "We never made a good living at all," he explains. "We'd have one or two years when it wasn't too bad, but the farm was going downhill structurally and I just didn't have the spare money to do repairs. It's terrible when you see your own farm going to rack and ruin around you."

Suicide rates amongst farmers are one of the highest of any profession, but Roger always found the strength to keep on going. "I never got so low that I got very depressed to the point of contemplating suicide – that's just not in my nature. I just kept fighting to come up with something to make it work. Turkey time was always good for us and Cynthia and I always tried to make the best of everything. We were desperate at times, but we stayed optimistic. We always had a lot of laughs, despite how bleak it got."

Then, one day, Cynthia had one of her lightbulb moments. She had noticed that despite there being a constant stream of visitors coming to Cannon Hall Park throughout the year, there were no public catering facilities in the parkland area. There had previously been a café at the museum at Cannon Hall, but for one reason or another it hadn't been a success and had closed down. Cynthia was convinced she and her friend Rosemary could do a better job and they suggested to Roger that they turn an old gun room at the farm, that was being used for storage, into a tearoom.

But to the Nicholsons' dismay, their plan almost fell at the first hurdle. Barnsley Council rejected their planning submission, in part, the Nicholsons were informed, because the council planned to re-open the museum cafe at Cannon Hall at some point. A local newspaper article covering the planning application included objections voiced by Cannon Hall Museum curator Brian Murray: "There is very serious concern about this application. There are very many people who would like to exploit Cannon Hall, but it is a sensitive area and will arouse public concern."

For the Nicholsons, having gone through the ordeal of the compulsory purchase order at Bank End, it felt like once again the council was laying their plans to waste. But as luck would have it, Harry Fish, the Mayor of Barnsley, just happened to live in Cawthorne Village and Roger invited him and members of the council planning team to Cannon Hall Farm to take a closer look at what the family was proposing. As they say, nothing ventured, nothing gained. "The mayor and the councillors came over to see the place and they were in one room and I was in the next and I heard Harry Fish say, 'Come on, give the man a chance'. He must have persuaded them because shortly afterwards we got permission to make it happen."

With plenty of manpower – and good friends to lend a hand – the wheels were set in motion to transform the old gun room into a traditional tearoom with waitress service and table space for 54 customers. "We looked in every direction for the best bargains and found classic cottage-style chairs and tables that we thought would fit the traditional style we were after," says Roger. "We've always believed that if you invest in good stuff it lasts longer, looks better and attracts more people in. It was a little dabble in a small business which set us off in the right way."

Meanwhile, Cynthia and Rosemary set about creating the tearoom's menu in the kitchen at the farmhouse. "Rosemary and I tried out all sorts of recipes and different ingredients until we found the perfect one for each item on the menu," Cynthia explains. "When we were perfecting our scones, we'd find that some were too flat and some were too soft, and one batch were more like rock cakes. Our friend Ken said that we should send them to the Falklands War to be used as weapons. I nearly threw one at him!"

The dynamic cookery duo went on to perfect the lightest short-bread, squidgiest flapjacks and chocolate cake, and rich cinnamon and almond slices. Roger willingly tasted every batch and was the official farm baking judge. "I think I put on three stone that spring!"

Rosemary's impressive meringues were an instant hit. Perhaps it was because her recipe called for no less than 10 egg whites. "We couldn't use eggs from the farm, though. The hens weren't able to lay them quickly enough!"

Finally, with all the recipes decided upon, teaspoons polished, menus typed out and Cynthia's mother Olive at the kitchen sink ready for washing-up duty, Home Farm tearoom was ready for its grand opening.

On 4 April 1981, the first customers were welcomed in and

the meringues, scones and other treats sold, well, like hot cakes. Everyone pulled together to help make a success of the day – even Roger had scrubbed his hands so he could help wait on the till. By the time the last customers had left and they could hang the closed sign on the door, everything had been eaten and there wasn't a tea cup that hadn't been used.

The family sat down together to reflect on the day's success. Despite having been worked off their feet, they had taken only £26, which was not enough for them to break even. "Well," said Roger, "that were bloody hard work and a right waste of time."

Having gone to all that trouble, though, they weren't about to give up. The very next day, the price list was tweaked: a glass of milk went up a penny, a cup of coffee with cream went up by three pence and so on. Meanwhile, a cream tea consisting of two scones, jam, cream, bread and butter and a pot of tea, cost the princely sum of £1.45 – surely a bargain in anyone's books?

Customers kept coming from far and wide and "Con" and "Flo" baked like demons all day. "I made fresh scones every day and would have one lot ready to go in the oven as soon as the other one came out," says Cynthia. "At one stage, I was baking 13 dozen scones every day. If anything ran out, we'd just bake more. And if something was really fiddly to make, we'd put the price up!"

To make sure nothing went to waste, Cynthia would use the yolks from Rosemary's massive meringues to make lemon ice cream. "It was just lemon, egg yolks, sugar and double cream and I'd mix it together and put it in the freezer. We wouldn't be able to make it these days, though, because of the health and safety issues of using raw eggs, but back then we just sort of did what we liked. To be honest, we just made it up as we went along."

For all the hard work – and it was very hard work just to break even – the two best friends had lots of laughs along the way. "I remember one time when we did a special salad tea for a group of old ladies from a nearby church group," Rosemary recalls. "Cynthia had terrible toothache that day and had a little bottle of whiskey in her pocket that she kept swilling around her mouth – for medicinal reasons, of course. Meanwhile, I was politely serving through the hatch when, all of a sudden, the church ladies started warbling a hymn. The two of us took one look at each other – Cynthia swigging from her whiskey and me at the hatch trying to keep a straight face — and we were near hysterical. Our shoulders were shaking and tears were running down our faces. We had to try so hard not to laugh."

Another time, Rosemary and Roger came back from the Cawthorne Carnival and, as usual, had gone to town with their costumes. Rosemary had been dressed as a punk rocker with a dress made out of a black dustbin bag, a dog collar and hair spiked up in various colours, while Roger, painted green from head to toe, was a very convincing Incredible Hulk. "Even though Roger had already had about three showers, he was still green and I just couldn't get the colours out of my hair, but we had to work in the tearoom," says Rosemary. "We were just getting everything organised when we heard the door of the team room opening and closing and a lady saying 'Mother! Mother! It's ok. I can assure you they don't usually look like this!'."

Home Farm tearoom certainly put Cannon Hall Farm on the map in the Barnsley area, but it took a great deal of hard work from everyone involved. As well as having to produce all the food, serve it and wash up afterwards, extra waitresses had to be paid, there were constant visits to the cash and carry for supplies, there was cleaning to do, daily cashing up and more expensive water and electricity bills to somehow pay for. It was a real struggle and it took three years before it started making any money. At that point Cynthia treated herself to a dishwasher: "For our house, you understand – not the tearoom – we had my mother!"

Always striving to offer more, the tearoom started to expand its menu with a bigger range of tempting treats. "We used to do toasted sandwiches, baked potatoes and things like that, then later on I started baking meat and potato pies, but it were mega hard work," says Cynthia. "I was making them by hand every evening for years and years. Eventually you realise you're better off getting in the right equipment and a professional to do it rather than trying to do it all by yourself."

The tearoom also grew out of its premises. Roger recalls one night where he had to take drastic action. "We had a party in the tearoom for a local artist and she'd invited lots of guests – more than we'd ever had in there before – and I was sure I saw the blimmin' floor start bowing. I ran to the woodshed and got long pieces of timber and started bracing the joists underneath with them which were bending dangerously. All the while I was sure I could hear the floor creaking and groaning. Thankfully it all held in place and the guests were none the wiser."

Knowing they needed to speculate to accumulate, it wasn't long until the family was thinking up plans for a bigger tearoom. But where would the money come from to expand? Plan C was needed.

Roger, Olive and Cynthia hard at work in the cafe kitchen

12

Opening to the public

The boys had all completed college. They had got their qualifications and the world was their oyster, but having been brought up at Cannon Hall Farm, they really wanted to help make it a success.

"It wasn't as if we all had a God-given right to stay on the farm," says Robert. The fact that we all loved living there was neither here nor there if we couldn't actually make it work, so we needed to do something fairly radical in order to survive."

"I remember asking all three boys what they wanted to do with their lives," says Cynthia, "and they told me they wanted to stay at home. So we had to try and make it a viable environment and create enough jobs for them."

"Back then," Roger adds, "I wasn't really interested in creating any more jobs than for the three of them. Well, that didn't quite go to plan!"

In 1938, London Zoo opened the world's first petting zoo, where people could safely have hands-on experiences with animals, petting them and sometimes hand feeding them. The idea proved hugely successful and the idea was taken up by several other zoos and, in turn, entrepreneurial farm owners who spotted an opportunity to diversify from being traditional food producers into tourist attractions.

And if other farms could do it, why not Cannon Hall Farm? While still operating as a working farm, breeding animals for meat and to sell for breeding stock, could the Nicholsons also open up their gates to the general public and have other animals on site that would interest them? With potential clientele on their doorstep from Cannon Hall Park, not to mention the constant flow of tearoom customers, the idea began to take shape.

The family went on various recces to other open farms in the area and also to bigger, more established businesses such as the Cotswold Farm Park and Chatsworth. They were impressed by how well organised each of the open farms were, if a little daunted by the amount of work that would obviously be needed to pull off the venture.

Many of the open farms operated as a very hands-on experience, allowing visitors to pet and feed the animals. Rosemary, never one to miss out on any fun, decided to give a billy goat a big cuddle, not realising that billies mark their goatees with their pungent urine to attract the ladies. Needless to say, the car journey home wasn't a pleasant experience for anyone. Another lesson learned.

With the experience of seeing other open farms in action, the Nicholsons knew that they would need to really start thinking out of the box if they were going to turn their farm into something that would be fit for purpose. Something radical was needed – this was not just going to be a matter of converting the odd building here and there, and even that was going to require serious amounts of cash. Which, needless to say, they didn't have.

Armed with a business plan that showed how some of the

buildings on their land could be sold for housing and others could be redeveloped for the visitor attraction, Roger applied to his bank for a loan, optimistic that they would buy into his bold new vision for Cannon Hall Farm.

The bank called in a special agricultural advisor to assess the plans and Roger was left in no doubt how they felt about the venture. "At the meeting they told me I'd never been able to support my wife and family and so I was better off just selling up. In other words, you've never been able to make anything else work, so why would this be any different?"

Things were bad, but Roger wasn't quite at the bankruptcy stage, as the farm was still worth more than the debt that he owed. The trouble was that the bank had extended Roger's loans many times before and just weren't confident that such a major diversification could actually work. "They just didn't believe it was possible to dig the business out of the hole that it was in," says Richard. "They thought the idea of opening up the farm to the public was just pie in the sky. They didn't take it seriously at all as a business proposition and the special agricultural advisor that they had brought in was equally unimpressed."

Far from stopping Roger in his tracks, the bank manager's words were like a red rag to a bull and galvanised him into action. There were plenty of other lenders out there and he wasn't going to give up until he had exhausted every option. In his heart, he knew that his business plan could work. If his old bank didn't want to know, Roger felt certain that there would be other investors who could see the farm's potential.

Thankfully, the family had the positivity of their accountant, Paul Howley, behind them. He'd worked with Roger for many years and believed the open farm idea was a winner. And with plenty of other contacts in the financial world, he was able to take the project to the manager of Yorkshire Bank, who immediately wanted to back the Nicholsons' venture.

Finally Roger got the funding to redevelop Cannon Hall Farm. His plans were ambitious, to say the least, and involved building a new house to accommodate all of the family and selling the original farmhouse and other buildings around the courtyard.

The first job was turning what was once a granary and grinding room, complete with the rumbling monster of a corn mill, into a family house. This was to be linked by a door to a granny flat for Rene.

The old farmhouse, which they were intending to sell first, was valued at around £200,000 and, along with the proceeds of the sale of the other buildings in the courtyard that the Nicholsons owned, the family intended to pay off some of their existing debt and invest a good chunk of money into the open farm project. They also applied for a small grant from the Ministry of Agriculture so they would have enough working capital to convert the farmyard and have all the necessary facilities for the visiting public. It was a very ambitious project and an awful lot could go wrong.

Roger was used to making business plans whenever he'd had to see if a farming project would be feasible, and the business acumen Robert had demonstrated as a child was obviously coming in handy. Not that Robert agrees. "We couldn't have been that good at business, though, as we never made any money! But it was a very tough environment – we were a small farm at a time when small farms just weren't prospering."

Back then, there just wasn't the appetite for niche farming that subsequently became so popular. Since the mid-1980s, the number of small family farms in England had started to drop significantly. Huge, industrialised farms put smaller homesteads out of business, as they simply couldn't compete with the cost of producing meat, dairy and crops on a small scale. If the Nicholsons were to survive, they needed a new strategy.

It was a case of evolve or die. "For us it was very much a siege mentality," says Robert. "We did a lot of sums and decided that everything that we could provide ourselves – such as some of the building materials and labour – we would do. We didn't go laying out money at all if we could help it.

"We'd always had food to eat, as we reared our own chickens and mum had her vegetable garden, so we didn't starve by any means, but we'd do whatever we could in order to stop money leaving us. We very much understood that money wasn't something that we were blessed with at the time, so in that way it wasn't a 'normal' existence. We certainly didn't have any money to chuck about, that's for sure. We had one car between us."

All those years ago when Charlie had taken a risk and bought the farm, rather than leasing somewhere, had really paid off. A fact that Roger couldn't have been more aware of: "If we'd been on a rented farm, we'd never have been able to do it because we wouldn't have been able to find the capital."

Roger's old friend from carnival days, Ron Carbutt, was integral in helping the family find a way of making the open farm idea a reality. As the managing director of a group of garages, he knew all the ins and outs about issues such as public liability and insurance, which of course was a whole new language for the Nicholsons.

With renewed vigour, enthusiasm and determination that this

project could actually work, plans were put in motion. Cynthia had very clear ideas about the layout and look of the new family home and Rene's flat, while the brothers set about clearing the other buildings ready for the valuers.

There was so much energy and enthusiasm between them. When the representative from the Ministry of Agriculture came to Cannon Hall Farm to advise the family about their new venture, he was bowled over by the range of talents the family had. After all, it's not many farms that have dedicated farmers, master bakers and artists all under one roof.

The target was to get everything done by Easter 1989, the busiest season for visitors at Cannon Hall Park. "We did as much of the labouring as we could, but we knew our limitations, so we didn't take on anything like the plumbing or electrics," says Richard. "Luckily, we didn't need to do an awful lot to the buildings that would be used for the animals. It was more a case of mending roofs, making them watertight and neater with new doors and boarding. Some of it wasn't of the highest standard, but we just couldn't afford to do that at the time. In any case, we always planned to upgrade the buildings in years to come if the open farm was a success."

Although the family had a decent budget to make the project work, they discovered early on that a huge proportion of their funds would be eaten up by one necessary, and very unglamorous addition, to the farm. One of the conditions of the planning application was that they installed a new septic tank and toilet system for the farm visitors. This involved the massive task of excavating a 20ft hole right through the middle of the farmyard with a 14ft drain from the house.

"It was an expensive job and threatened to take all our cash," says

Roger. "But letting our contractor go and accepting help from my brother-in-law saved the day."

As luck would have it, the boys' cousin Alwyn (Shirley and Laurie's son) just happened to have access to a digger, as he previously did landscaping work for the National Coal Board. It wasn't exactly state-of-the-art technology, but it got the job done. It was a mammoth task and with the cost of getting the professionals to lay pipework and build the tank itself, it wasn't cheap either. You didn't need a degree in accounting to work out that there would be very little budget left for anything else.

That made it even more important for Richard, Robert and David to pitch in with the labouring; dismantling roofs, patching up buildings and laying concrete floors. It was hard, physical graft and it didn't help that it had to be done in the middle of winter, but with the deadline of Easter 1989 looming, the clock was ticking.

As the rain lashed down and the pile of earth got bigger and bigger as the septic tank was created, it was hard to keep positive about the task ahead. Having had to fork out so much money for what was essentially a massive hole in the middle of the farmyard was dispiriting to say the least. The family had started the project with technicolour dreams about how fantastic the open farm could be, but with their budget drastically adjusted, some of their more ambitious plans had to be put on the back burner – for the time being at any rate.

13

Animal magic

Building work aside, there was also the small matter of acquiring interesting animal attractions that would draw in the crowds. Herds of fluffy sheep and healthy cows look lovely in the fields in any weather, and a litter of snuffling piglets is always very cute, but the Nicholsons knew they would have to up their game if they wanted people to actually stump up money to come and visit.

"I remember, when I was a kid, I drew a picture of all the different kinds of animals we could have on the farm," says Robert. "I had pens for rabbits here, llamas there, snakes and all sorts of exotic animals. Now we were actually going to make that happen."

While the brothers' imaginations went into overdrive, Roger was always keen that the Open Farm should show off the traditional side of farming; he also recognised the educational potential for school groups. As the opening was scheduled for the Easter season, there would also be plenty of newborn lambs and calves being born and lots of newly hatched fluffy chicks to pet.

Knowing there needed to be more than conventional farm animals, Roger also invested in the farm's very first Shetland pony (many more would follow over the years, including the famous

Jon Bon Pony) and he also gave a new home to two Vietnamese pot-bellied pigs called Maggie and Denis. "I suppose it was a bit strange going from buying traditional farm animals to things like pot-bellied pigs," says Roger. "But it was exciting to be trying something new – and another part of the challenge to try and get it right. I've always enjoyed going to market and buying animals, so it didn't seem weird at all – just enjoyable. There were things we had to do a bit of research on before we bought them – and along the line we've certainly learned from buying badly – but back then it was all very exciting."

"We borrowed some Angora goats, which were very trendy at the time," says Richard. "A friend had brought them over from Europe with a view to making some money on them, only it never quite happened. They were very cute, though. Mum also had her eye on some Angora rabbits, with the idea that she'd get wool spun from the fibres they shed, knit it into jumpers and sell them."

Then there was Victor the llama. "We were sold him with the story that he'd appeared in the *Doctor Doolittle* movie," says Richard, "but in hindsight every llama in the country seems to have been sold on that basis! I don't think it is him. The one in the pictures with Rex Harrison looked nothing like our Victor."

For several weeks, the boys pulled in favours, rang around for good deals and put feelers out for new animals that could be housed in the new, improved farmyard. With the final push to get everything finished – and the small matter of Robert and Julie's wedding just a week before the farm was due to open, everyone was on tenterhooks. This really was make-or-break time.

"We knew we were in the last chance saloon and that this was the last throw of the dice," says Robert. "If this venture failed then we were pretty much done in. The other problem at that stage was interest rates were going through the roof and we just managed to sell the last house in time to take us debt-free. If we hadn't had a sale, I'm not sure that things would have worked out as they did. We may not have survived."

Just one week before the official opening, Robert and Julie tied the knot at St Patrick's Church in a village near Julie's home town of Bedale in front of 100 guests. It was followed by a huge party in the evening at a hotel near Catterick racecourse. "We had a couple of nights away as a honeymoon in Newcastle and then we were straight back to work," says Robert. "Julie was working in the cafe and I was helping to get the farm opened. It wasn't exactly the holiday of a lifetime!"

After one last push to get the farm ready for the public - putting up Entrance and Exit signs, removing litter and making sure all the windows were gleaming, the family prepared for the next milestone in their life.

On Good Friday, 24 March 1989, Cannon Hall Farm and Tearooms officially opened to the public. The question was, would anyone want to come and visit?

A couple of local newspapers had run stories about its opening and friends had been putting the word round, so by the end of the first day around 50 customers had come through the gates. Not great numbers, true, but it was a start. A few more came on the Saturday, then around 100 on Easter Sunday and maybe 150 on Bank Holiday Monday. The numbers were going up, which was the important thing, and with admission at £1 per adult and 75 pence per child, it meant there was money coming into the farm rather than going out of it for a change. "We were really happy with that and it totally exceeded our expectations," says Robert. "We all thought it was very good considering what we were offering at the time."

Like all new ventures, there were teething problems and once again Roger found himself on a steep learning curve. "The first day we opened I was a bit embarrassed," he says. "Although the place looked nice and clean, people were pushing buggies and prams over limestone chipping and bumping along and it wasn't exactly a smooth ride. But tarmac was expensive at the time and we were venturing into foreign waters knowing what we could afford to do."

A few lumps and bumps here and there certainly didn't stop the visitors from coming and throughout that first summer, more and more customers excitedly lined up to pay the admission price. Within just a few months of being open, the brothers were looking to add to the number of animals on the farm to make sure people didn't go home disappointed. Rheas, emus and wallabies were soon added to the line-up, along with a miniature Dexter cow and her tiny five-week-old calf, Sprogg.

Meanwhile, rodent lovers could head to Mousetown, a series of tanks designed to house mice and rats, with a hand-painted backdrop created by resident family artist, Richard Nicholson.

With lots of baby animals to be seen and petted, it wasn't long until school groups were booking in to visit the farm for its educational activities. Roger, Richard, Robert and David devised a tour around the farmyard, a show-and-tell about all the animals, answering questions and bringing the world of a working farm to life. Did you know, for instance, that a pig is pregnant for three months, three weeks and three days? Who needs a theme park when you have edutainment like this on your doorstep!

"We've always got an immense amount of pleasure educating children and showing what we do on the farm," says Roger. "And from the word go they've seemed to enjoy it, too. Things were looking up for us and we were all experiencing that excitement that I had when I first moved to the farm. It was such an optimistic time for us all."

The news of Cannon Hall Farm's Dexter calf Sprogg made it into the local paper and soon other stories from Cannon Hall Farm were hitting the headlines. "One of our earliest ruses was to name the animals after *Coronation Street* characters," says Richard. "We had Tanya the Burmese Python who was featured in *Chat* magazine – she was named after a barmaid at the Rovers – and then we had two rheas called Martin and Gail, named after the Platts, and a pig called Reg Holdsworth. One magazine ran the headline 'Reg Holdsworth is a big fat pig!'"

As anyone who has ever been on a school trip will agree, one of the best things about the day is hitting the gift shop at whichever stately home, zoo or other attraction you are visiting. So it wasn't long until the subject of a gift shop came up at Cannon Hall Farm.

Unfortunately for Richard, however, this was one idea too far for Roger. "I remember suggesting we had a few pencils and rubbers around the cash desk with Cannon Hall Farm printed on them," says Richard. "But my dad said, 'What would people want them for?' And when I told him I'd probably have to order 1000 of them in one go, he said, 'it'll take us two years to shift those!' I was hardly being Richard Branson, but I could see the benefit of having a few souvenirs for children. I understood that not everything would sell, but you just have to try things. It's always going to be a bit of trial and error."

As more and more visitors came to the farm, Richard managed to convince Roger that a gift shop could be a good idea. Cynthia was true to her word and knitted up some Angora rabbit jumpers and also on sale were candy canes, his 'n' hers cow mugs (illustrated by Richard) and handmade wooden toys. It wasn't a huge selection, but it made the place feel like a proper attraction – and of course the extra takings came in very handy.

"We had a big old-fashioned cash register that was in 'ds' – not pence – that's how ancient it was," says Richard. "It had a clunky pull-down lever and you had to get out of the way when the drawer opened or you'd do yourself an injury. We hadn't the money to splurge on anything at the time, so we tried to make a feature of how antiquated it was, even though it was actually a bit embarrassing that we didn't have a proper till!"

Having to balance the books after their massive initial outlay meant that the first year of running an open farm was certainly a challenge. Nevertheless, at the end of the first season, which ran from Easter until mid-September, some 13,000 visitors had come through the gates of Cannon Hall Farm. Things were definitely looking up.

"The more people that kept on visiting, the better," says Robert. "After all, they were the people that were allowing us to stay living there."

1 4

New arrivals and red tape

In order to encourage new customers and past visitors to make return visits, the Nicholsons realised that new attractions were needed. Over the course of the next few years, the family introduced a number of new breeds to their Barnsley farm – and also welcomed back some old favourites.

"We'd seen pictures of Dad on his old shire horse Blossom," says Richard "and we thought it would be great to have another one back on the farm. But Dad wasn't keen initially as they are expensive to keep."

Because of their size, shire horses cost more to feed than smaller horses and are more prone to certain equine diseases. But fate was on their side as the brothers' cousin Richard (Alwyn the digger's twin brother) owned shires and was happy for one to be loaned to the farm.

Her name was Blossom – the same name as the shire Roger had loved as a child – so it felt like fate that she found her way there. And like so many of the four-legged favourites at Cannon Hall Farm, she was the first of many rare breeds to be welcomed into the family.

To coincide with the opening of the farm at Easter 1989, Roger invested the rather steep sum of £50 for a chubby pygmy goat called

Carla from a local seller. Once again, the name Cannon Hall Farm ended up in the press when Roger got more than he bargained for. "Sometimes pygmy goats have pot bellies," he explained, "and Carla's owner had assured me that she wasn't pregnant. However, six days later, she had two nanny kids. So I ended up with three goats for the price of one! I was delighted as money was quite tight back then."

The two pygmy kids were named Dandelion and Burdock and, like Blossom, were the first of many of their breed to make their home there.

With more visitors came more vehicles and soon the unexciting but necessary issue of car parking came up. "We used to let coaches park on the field, but we got a telling off from the council," says Cynthia. "We understood you could more or less do what you wanted on your own land for 28 days, so that's what we had been doing – only we'd lost count... but the council hadn't."

Once again, the family had to decipher a deluge of paperwork to satisfy the powers that be. "To justify building a car park," says Richard, "we had to collect a dossier of how many people were parking on the roadside and on public land, so we could show there wasn't adequate space for them and that we really needed to have our own car park."

As well as having to painstakingly count the number of cars coming to the farm every day, the family needed to do more than just create a car park. Because any development on their land would be able to be seen from the surrounding area, the car park couldn't be a blot on the landscape and had to be sympathetic with the environment. A great big concrete slab with painted rectangular spaces for cars isn't exactly a pastoral tableau, so the Nicholsons commissioned a landscape architect to design a different kind of car park – one that they hoped would aesthetically fit with its surroundings.

Incorporating natural elements such as trees and shrubs, the car park design certainly fulfilled the brief. However, the best position for its location was Cynthia's beloved vegetable garden. She wasn't exactly thrilled with the idea – possibly the understatement of the century – but she understood how necessary the new car park would be, so construction (and destruction) went ahead.

However, there was a silver lining, when the new car park went on to win a design award – one of many to grace the mantelpieces at Cannon Hall Farm. Says Roger: "It was a very demanding time for all of us – not least because Cynthia had to lose her garden, which she really loved. But it was a real time of growth, so it was very exciting – even if it was just a car park. Nonetheless we had to balance how much we could afford to do with how much we were bringing in."

The Nicholsons then had to tackle the issue of which kinds of animals were technically farm animals and which were zoo animals. Llamas obviously weren't as common as cows, sheep and pigs, but they ticked the farm animal box. If, on the other hand, they wanted to up the number of more exotic breeds such as snakes and tropical fish, the family would have needed to apply for an expensive zoo licence, which at that stage was a step too far. As much as they would have liked to have more weird and wonderful creatures, there were other priorities.

Knowing how important it was to keep children entertained,

in 1992 the Nicholsons decided to take the plunge and create an adventure playground on the farm. "In the grand scheme of things, they seem very expensive for what you get," says Roger. "But you have to comply with all the health and safety regulations, which was another new language for us to learn."

With the price tag coming in at tens of thousands of pounds for even the most basic playground equipment, the family had to weigh up whether an adventure playground would be used enough to justify the expense. But from the day it opened to the enthusiastic public it was obvious that the new attraction was going to be well used and enjoyed.

And with more and more visitors to cater for, the tearoom was always busy and the gift shop was nearly selling out of those Cannon Hall Farm rubbers and pencils. Richard had proved his dad wrong – people do love a branded souvenir. So it wasn't too long before plans were underway to upgrade both the tearoom and gift shop. Cannon Hall Farm was growing at quite a pace!

A brand new building was constructed as an expanded tearoom for 100 customers, plus a bigger gift shop for souvenirs, including personalised pop-a-point pencils and jars of Cynthia's fantastic homemade plum jam. It would be a while before the brothers' catch-phrases "Oh 'Eck" and "Happy Days" would make an appearance on T-shirts and aprons, but the till kept ringing, so the brothers knew they must be doing something right.

The new tearoom/gift shop was built in such a way that extensions for more units could be added on, should the need arise. Think man size Lego. And judging by how much the farm had already grown, it was more than likely that there would be more additions.

With the old tearoom now standing empty, there was a way for

David to move back into the farm again. Although new buildings had been constructed on the farm, he and Anita weren't able to build a conventional new house there. They tried everything from a log cabin to another barn conversion. But plans to build more human housing from scratch were rejected and he and Anita had been living in the centre of Barnsley. As a result, David had been travelling back and forth to work every day. In the usual scheme of things that would be fine, but when you work on a farm it's much more practical to be on site at all times – especially when there's an emergency in the middle of the night and an animal's life depends on you being there.

Converting the tearoom into a house for David and Anita was a fantastic solution and, after a tricky conversion – removing the original horsehair plaster nearly brought a supporting wall down – their new pied-a-terre was ready. Their living room was the former tearoom and the gift shop downstairs was turned into bedrooms – a perfect upside-down house. "It worked really well," says David, "the only trouble was some people still thought it was the old tearoom, so we'd get people walking into the house asking for a cream tea and a slice of cake."

It was taking the term open farm to a whole new level...

"We never set out to be a huge farm," says Roger. "We've just gradually grown and grown. And as we've evolved, we've discovered you need highly qualified people to make things work. In the beginning it was all a bit 'make-do-and-mend' with Cynthia and Rosemary running the cafe and family members helping out. But when you have a proper cafe, you need a professional chef who knows what they're on about. It's no good just having someone who can fry a few chips!"

A bigger venture meant everything had to be scaled up, from the industrial food mixer to a more sophisticated coffee machine. At that stage, Cynthia was still baking every day and working in the cafe full time and more staff were taken on to help her.

"When the new tearoom opened, I signed over my share to Robert's wife Julie," says Rosemary. "When it had just been me, Cynthia and Rene it was a different ball game, but now it was a much bigger project – and a real family business. I was relieved, though, to be honest. I couldn't face toasting another bloody teacake!"

(Rosemary fans will be happy to know that she is still very much part of the farm and now lives in the beautiful Tower Cottage.)

15

New challenges

J ust when everything was ticking along nicely, stories of Foot and Mouth Disease started hitting the newspaper headlines.

In February 2001, the first cases were detected at an abattoir in Essex and pigs from Buckinghamshire and the Isle of Wight had both tested positive with the disease. Later in the month, more infected animals were discovered in Northumberland, Devon and North Wales. By the beginning of March, the disease was spreading like wildfire through the Lake District.

The outbreak caused a massive crisis in both agriculture and tourism as 2,000 cases went on to be identified. Tragically, more than six million cows and sheep had to be killed in order to halt the disease.

While some farms lost everything, with all of their livestock having to be destroyed, fortunately Cannon Hall Farm was spared, as none of its animals tested positive and the farm received a clean bill of health. But the incident was a wake-up call that no matter how much you plan ahead and try to be prepared for every eventuality, sometimes fate has other ideas...

Closed signs were put up at the farm's entrance and remained there for three months. It was over the Easter period, which normally would have been the farm's busiest time, but the family had no choice but to shut up shop. "It was absolutely tragic for the people who were affected and had to destroy their cattle," Roger recalls. "Some people lost their whole livelihoods overnight. There weren't that many open farms back then and even though humans can't catch the disease from animals, we felt it would be irresponsible if we didn't shut down."

If nothing else, it allowed the Nicholsons to press pause on their non-stop busy life and enjoy spending more time with the next generation of the family. By that time, Robert and Julie's children, Tom and Katie, were 11 and eight, and David and Anita's daughter Poppy was six. For 12 weeks they had free run of the farm.

While the farm had been growing around them, Roger's grandchildren had been growing up, too, surrounded by animals just like generations of the family had before them.

Luckily, once the foot and mouth episode was over and farms were considered safe again for tourists, it didn't take long before visitors were flooding back and enjoying the new, improved facilities. And never ones to let the grass grow under their feet, the Nicholsons were planning the next new venture for the farm.

"For a while we'd been talking about selling our own produce through the farm shop," says Richard. "The prices that we were getting for our livestock at the market were really low, so we decided the only way to make it work was to cut out the middleman and sell directly to the public."

To try the idea out, the brothers got in touch with Alan Asquith, the butcher from their Aunty Shirley's farm shop, and he prepared a selection of cuts to be sold in their freezer in the farm shop. "We didn't try hard with the packaging – it was a bit rough and ready – but the quality of the meat was good," says Richard. "So we put a sign on the door saying 'Meat for sale in the deep freeze' and low and behold people started buying it. We did remarkably well considering it was just a domestic freezer, not a proper display cabinet."

Considering the family did nothing to publicise the farm shop, more and more customers started to visit to buy meat and the Nicholsons realised that the idea had legs. "With our *Field of Dreams* mentality, we just thought, if we build it, they will come and we decided to take the idea further."

The brothers were introduced to a butcher called John Holmes via 'Chicken Jim' Wragg, the family's chicken supplier. John was in the process of shutting down his Sheffield business and was looking for a new venture. "We were very lucky that we found him because he set us on the right track with how to sell meat because we really didn't have a clue," Richard explains. "He taught us so much and Robert and I would work with him in the butchers shop making the sausages and it was really good fun."

With word-of-mouth success and repeat business, it wasn't long until more hands were needed on deck. The boys' Aunty Shirley was closing her farm shop, so her butcher Alan came to work at Cannon Hall Farm. "He was the one who chopped up our very first meat and he's been with us ever since. He's a real hit with the customers."

Next came the decision to create the farm delicatessen. This would be the perfect way to showcase local produce, such as fantastic cheeses and cured meats and the family could also produce cakes and pastries from the deli's bake room – including Cynthia's original recipe for scones, now made famous as "The scones that saved the farm".

Like so many episodes in the story of Cannon Hall Farm, the right people seemed to come along at just the right time. Two weeks before the deli officially opened, a couple of local pie makers called Rob and Christine Webb came through the doors. "They'd been selling pies at Penistone market," says Richard. "But they wanted a new challenge in their lives and asked if they could come and work for us."

One bite of their excellent pork pie later, and they were joining the team.

Slowly and surely everything was building up nicely. "Well, it wasn't that nice," Richard laughs. "We had a horrible green carpet on the floor for many years because we were too tight to change it. It was one of the floors that Rob, Dave and I had laid so it was very uneven. But at that stage it would have cost too much to level it out!"

It was becoming obvious to everyone who was now involved in the farm that even with all the fairly recent expansion, it was quickly growing out of its existing premises. The family could never have predicted what a success story the venture would have turned out to be. From being just a humble homestead, it was now a major Yorkshire tourist attraction.

"It felt as though the farm had grown old around us," Robert says. "We knew that to invigorate the business we had to invest in it and by building a bigger new farm we would be safeguarding our future. We wanted to improve all of our existing facilities and create modern, well designed paths that visitors could walk around without fear of treading in cow or sheep muck."

In 2009, the Nicholsons worked with a major architect to create a five-year plan to future-proof the business. It included an upgrade on everything from the toilets and tearooms to new farm buildings including a state-of-the-art circular cattle-handling system called a roundhouse, complete with a visitors' viewing area. There would be a new indoor play centre (for those few – and very rare! – days the sun doesn't shine in Yorkshire), a new restaurant and bar, and a new and improved farm shop showcasing the best produce from the farm and local suppliers.

"We knew it was a very bold move," says Robert. "But after the financial crash of 2008 there were a lot fewer planning applications coming into Barnsley Council, therefore our application would liter-ally keep someone in a job. We knew it was a big step coming out of a recession, but we were confident that we had a good few years ahead of us and recognised we had to spend real money in order to be able to properly reinvigorate the business and set ourselves up for bigger and better things."

Once again, a massive sum of money had to be allocated for perhaps the least interesting area of the farm – the toilets. "We knew that the toilets weren't up to spec for the amount of people that we hoped would visit once we had renovated," says Richard. "But it took a fair chunk of money. People don't think that when you are opening a visitors' attraction that the toilets will cost over £100,000."

It wasn't going to be possible to make all the changes in one hit – the Nicholsons hadn't suddenly become tycoons overnight – but by developing a five-year plan, they could map out their priorities and move ahead as and when they were able to build more.

Once the toilets were looking shipshape, it was time to take on the most exciting addition to the farm to date, the Roundhouse. The new circular structure was topped with a unique fabric roof, similar to the one used on London's Millennium Dome, and was designed to house around 100 cattle. It was open on all sides to allow easy entry and exit from each of the pens and would have a viewing gantry reached by stairs. It would be five-star accommodation for any four-legged guest and a great place for visitors to come and watch farm work in action.

This time around, the brothers may not have been doing as much of the physical labouring as before, but with so much experience behind them, they were able to project-manage the upgrade. They were in the unique position of knowing exactly what was needed for farming purposes and also what would work as a visitor attraction.

"We all loved the design of the Roundhouse," says Robert. "The fabric roof was made in Finland and it was lowered down on to its supporting poles, it looked like a giant flying saucer. I did worry that it might be a bit of a gimmick and the roof might not be strong enough for Yorkshire winters, but we were happily proved wrong."

And from the very first day it was opened, the Roundhouse has proved to be one of the most popular locations at Cannon Hall Farm, both for its birds-eye view of the animals on the ground floor, but also as a shelter when the weather misbehaves.

Things were going from strength to strength. But, as is always the way, once the brand spanking new Roundhouse was in place, the older buildings started to look a bit shabby. After a couple of years, it was time to call in the diggers again.

16

Time to think big

The Nicholsons' five-year plan for Cannon Hall Farm had been drawn up in 2009 and many of the expansion projects had been put in place. But it was in the years 2013-2017 that the farm was really transformed.

"One of the biggest mistakes we've made as a family is actually not thinking big enough," says Robert. "When you are starting out, you can only build what you can afford, but having a grand plan drawn up definitely gave us something to work towards."

Before the major conversion, buildings had been patched up, repurposed and more or less assembled when they were needed. But once the spaceship-like Roundhouse landed, the Nicholsons knew it was time to think big. Cannon Hall Open Farm had been such a success that they could embark on a £1.4 million development to totally transform the farmyard.

But it wouldn't happen overnight. In fact, it took nearly a year to erect seven new farm buildings, each with a purpose-built viewing gallery. One was built as a milking parlour and the others would be farrowing houses for pigs and barns for rare breeds.

From the word go, the Nicholson family had always championed less common animals. Welcoming shire horses to Cannon Hall Farm early on was a great way to spread the word about their scarcity, and over the years the farm has introduced all sorts of rare animals to the public including Swiss Valais, Soay, Jacob and Herwick sheep. "When a breed dies out, that's it. It's gone and you'll never get it back," says David. "If we can do our bit to safeguard even one rare breed then it's giving something back to the countryside and the heritage of farming."

In 2009, the farming community was again in the headlines when an outbreak of E. coli disease infected visitors at a petting farm in Surrey. The news sent shockwaves through rural tourism, as being able to touch and stroke the animals was one of the reasons open farms had become so successful.

Petting zoos were said to be a breeding ground for drug-resistant superbugs that were particularly dangerous to small children. The disease could spell the end for small petting farms and could drastically affect the numbers of visitors to Cannon Hall Farm. "Being able to touch the animals was part of the educational benefits to children," says Roger. "But there's no way we could inflict the potential of any illness on them, so we decided to take any risk out of it and change the way we operated."

Some animals, such as rabbits, carry a very low E. coli risk, so the decision was made to keep them in a supervised petting area with lots of washing facilities nearby. They also devised feeding chutes next to some of the animals, so that children could still experience the fun of feeding the animals. And extra barriers between the animals and visitors were set up to further reduce the E. coli risk.

With fewer opportunities to get hands-on with the animals, but obviously still wanting children to have plenty to keep them entertained, it was time to upgrade the play areas. Unlike the last adventure playground, which had all the usual playground favour-ites such as slides, swings and plenty to climb, the new improved adventure playground ramped up the fun factor with a giant tube maze, fantastic tower slides, a climbing forest and even a zip line. The only trouble was, when the weather was bad, it wasn't much of a draw.

Step forward The Hungry Llama. In Spring 2015, the new indoor play area opened with a whole new world of exciting bore-dom-busters for little ones. It was attached to a vast new 400-seater family restaurant and its name The Hungry Llama was inspired by Victor, Cannon Hall Farm's first ever llama – who always seemed to be hungry!

Having had a tearoom at Cannon Hall Farm for over 30 years, the Nicholsons decided the time was right to create a much bigger restaurant that would also have a licensed bar.

All that was needed was a top chef to make the magic happen – and in true Cannon Hall Farm style, the right person for the job just happened to be available. *Great British Menu* celebrity chef Tim Bilton had often bought meat from the farm for his excellent restau-rant, and one day while picking up supplies at the farm shop, he heard about the new restaurant venture. The timing couldn't have been more perfect as he was looking for his next challenge. The idea of creating a farm-to-fork menu got everyone's taste buds tingling.

With Tim Bilton at the helm, it wasn't long before more appetising ideas were being cooked up for foodie fans in Yorkshire, including bistro nights, cookery demonstrations and even a food festival.

And once the Cannon Hall Farm website was getting more and more virtual visits, Tim's fantastic recipes could be shared online. Everyone agreed that the restaurant should be called The White Bull, after Charlie Nicholson's famous beer-drinking shorthorn bull.

And human foodies weren't the only ones to be served up a treat at Cannon Hall Farm. Having had many enquiries whether their restaurants were dog friendly, it was time to go one better – with a dog friendly cafe that even had a menu for dogs. Puppuccino, anyone?

It seemed fitting that in the traditional way of naming the restaurants as a tribute to Cannon Hall Farm heroes, the new venue celebrated Roger's old sheepdog Flossie. Back in the days when Roger used to farm with his friend Will Roe, Will's sheepdog Lassie would often accompany him. Later, when Will retired, he asked Roger to keep Lassie as she was very much a farm dog and he knew she'd be happier there. Lassie went on to have a litter, then two days later, two more pups arrived, but one was stillborn. The other was hanging on by a thread but in true Roger style he wanted to give the young pup the best chance in life. He decided to keep the lucky pup for himself and named her Flossie. "I know I'm a softy," he says, "but we often keep animals on the grounds that they've had a little incident in their lives and they need extra help."

By all accounts, Flossie never quite got the hang of being a sheep dog and Roger would be hoarse after a day trying to round up his flock. Even so, little Flossie is a part of Cannon Hall Farm history and is captured with Roger on the official farm logo.

In 2016, Roger was going about his work on the farm as normal when one day he had a nasty accident involving a 600-kilo cow. The cow was lame and Roger, thinking that it might have an infection in its foot, got it into the cattle crush examination cage so he could investigate further. He reckoned it was a relatively simple matter of cleaning up the cow's foot to get a closer look, then giving it a spray of antibiotics.

Roger opened up the side door of the crush so that he could get a hosepipe near the cow's leg to give it a good wash. Everything was going fine, until he squirted the antibiotic spray and it gave the cow such a start that it kicked out at the cage door, which rebounded into Roger with an almighty thwack. "I remember standing there for a few minutes and thinking, 'What happened there?' and one of the farm lads telling me to sit down while he rang home to get help. I don't remember being in agonising pain, but something definitely felt amiss and by the time I got back home I was as white as a sheet with shock."

Knowing that an ambulance was on its way to the farm and that he'd be going to hospital, Roger decided he needed a bath first as he was covered in cow muck. But halfway through getting undressed, he realised he wasn't up to it. "I just wanted to lie on the settee and have a rest. By this time there seemed to be loads of people in the house – who knows, maybe they'd come to wave me goodbye? I remember drifting off to sleep. Everyone was shouting at me to stay awake. Then I was stretchered off to hospital in Sheffield – those ambulances are very bumpy! I was x-rayed – I'd broken four ribs and I'd punctured my lung."

There's not that much that can be done with a broken rib injury and once Roger's lung was re-inflated, after four days he was allowed back to the farm – and ordered to rest. But like his father before him, Roger wasn't very good at taking it easy – especially when there's work to be done.

"If I have to go to hospital that's a different thing," says Roger. "I'm not a bad patient because I actually like hospital food and I do what I'm told, because that way the sooner you start to get better, the sooner you can get out. But even though when I got home I couldn't do any major farm work for around two months, I still walked around the farm and checked on things. I'm never completely out of action."

Roger's accident was a reminder that farming can be a very dangerous job and that no matter how many times you might do something very routine – like giving a cow a bit of first aid – animals can be unpredictable, especially when they weigh over half a tonne.

David also remembers having a close shave when a grumpy cow landed a kick right into his chest. He fell to his knees, turned purple and couldn't breathe, but thankfully he was fine once the shock wore off.

Older brother Robert was less lucky when he was moving cattle one day and his finger was nearly severed by a metal chain. "I remember thinking, 'Oh 'eck, that's serious' and I held the parts of my finger together while David drove me to hospital. I was really worried as I had cow muck in it and it was bound to get infected, plus I had to hang around for ages before I could be seen by a doctor. Then I had to have emergency surgery to re-stitch the nerves in my finger before I was allowed to go home." Even though Robert now had a new and improved bionic finger, it took months for it to heal properly and he had to go through weeks of physio and rehab.

17

The wild and wonderful

Since its opening day in 1989, Cannon Hall Farm had gone from being a small local farm with a few animals to visit and pet, to a world of wonderful breeds to discover, the best adventure playground for miles, several superb restaurants and a fantastic farm shop.

By 2017 it was time for a re-think about getting a zoo licence. "When we first opened the farm to the public, the local authority gave us special dispensation," Robert explains. "But once you get to a collection of a certain size, it triggers the need for a full licence."

In summer 2017, the new reptile house opened and the number of animals multiplied considerably, so the zoo license was very definitely needed.

"In order for a zoo license to be granted, everything has to meet strict criteria," Farmer Ruth explains. "Everything has to be spot on, from the animal welfare and husbandry to health and safety and all of the paperwork records." With Ruth in charge the inspector gave the farm a year to get everything ship shape.

"We found most of the reptiles at a West Yorkshire charity called Reptilia, that had rescued the animals from all sorts of unsuitable domestic settings," says Robert. "We bought whatever they had in stock because we needed things to start up the new collection – so that was newts, crabs, chameleons, a python and a four-foot-long female monitor lizard called Rex – who still happily lives at the farm now."

And as with all new guests at the farm, the reptile house was kitted out with ideal habitats for them to thrive, including roomy heated tanks and a back-up generator in case of power-cuts. Staff were trained how to handle creatures for the show-and-tell sessions – and more importantly, which ones to avoid cuddling up to. When some of the more dangerous creatures arrived, they were left to settle in on their own terms...

With the addition of the Reptile House and all of the other new facilities, a whole raft of specialised staff needed to be employed to keep everything running smoothly. A full-time office team was put in place to deal with enquiries from the visitors, apprentice chefs and trainees were taken on for the new restaurants and a whole army of behind-the-scenes experts ensured the farm's smooth running.

In spring 2014, to coincide with the 25th anniversary of the farm opening to the public, the Nicholsons decided to make the entry price to the farm £1 for adults and 75 pence for children – the same they'd charged on the very first day of Cannon Hall Farm being opened to the public. The idea was to celebrate how far they had come and potentially invite a few more visitors in.

Unfortunately, the idea backfired. "We chose a random week day, not thinking that anyone would particularly take advantage of it," says Richard. "But that day we were absolutely rammed. People from miles around took their children out of school for the day and the roads were gridlocked with traffic!"

Cannon Hall Farm was now such an established part of Barnsley life that the local newspapers and Yorkshire's biggest regional newspaper *The Yorkshire Post* often featured stories about it, from unexpected arrivals and new breeds, to the special events that were regularly being held there.

And it wasn't just the facilities and activities that were getting more media attention. In 2015, the Nicholsons received a royal visit in recognition of employing 250 full-time members of staff, therefore contributing significantly to the Barnsley economy. With rising unemployment rates throughout the UK, Cannon Hall Farm was bucking the trend by being able to employ more and more people on the farm.

On 15 May, Prince Edward, the Earl of Wessex, spent a morning being shown around all the new developments and meeting new members of staff. He even bet on the winning ram William the Conqueror at the sheep racing track. Roger found pictures of the Queen and Prince Philip when they had visited in the Jubilee year and showed them to Prince Edward, who said, "Oh look! Mummy and Daddy were here!"

Not surprisingly, the royal visit made the papers and the local television news, but by then the Nicholsons were used to seeing themselves on the small screen. For a number of years, during springtime, local television news teams had filmed stories about the lambing season. They had also worked hard at establishing a strong social media following.

In 2017, the Nicholsons got into the final of the National Farmers Weekly Awards in the Diversification Farmer of the Year category and a BBC television production team followed them to the glitzy grand finals in London. Unfortunately, they didn't win the award – nor the following year when they were shortlisted again. Once again they were accompanied by a film crew but didn't take the prize back to Barnsley. "We were becoming experts in failure!" Robert laughs.

But somewhere along the line they must have shown that they made good television…

18

Springtime on the Farm

In 2017, Daisybeck Studios, the award-winning production company behind the Channel 5 series *The Yorkshire Vet*, about Thirsk vets Peter Wright and Julian Norton, got in touch with the Nicholsons about taking part in a new television programme called *Springtime on the Farm*.

As its name suggested, it would follow the events unfolding at a farm in Britain during lambing season and the plan was for the show to air from Monday to Friday over the course of one week, with two presenters based at a host farm and various celebrities popping in occasionally with special reports.

"The head of the local tourism agency Welcome to Yorkshire had recommended us to the production company," says Robert. "At that stage, they didn't know what the format of the show would be, except that it would be five nights in a row based around lambing."

Knowing the programme could be great publicity for Cannon Hall Farm, the boys pitched hard to become the chosen farm. "Initially, we were a bit worried that we would have to provide all the content in the show," admits Robert, "so we were relieved when we heard that there would be other reports from around the country."

"We were looking for a host farm that gave us a bit of everything – a family-run farm where the animals were the focus, but also with a great story to tell," says Paul Stead, the Managing Director of Daisybeck Studios. "My wife had visited Cannon Hall Farm for years with my children and she'd told me how fabulous it was, so I arranged to meet the three brothers and Roger."

"Cannon Hall Farm is like Disneyland in terms of its professional set-up," says Kim Metcalf from Daisybeck, who has worked so closely with the Nicholsons that she even has a donkey named after her. "When the team was looking for a farm where they could base the programme, and a farming family that would fit the bill, they couldn't help but be impressed by Cannon Hall Farm and the Nicholsons.

"As soon as I met the Nicholson family I knew Cannon Hall was the perfect place. It was the first farm that we visited to consider for the programme, but they set the benchmark so high that no other farm could offer us anything more," says Paul.

"Geographically it was spot on as it's right in the middle of England and as we had a large crew at that time it was the ideal set up, because the Nicholsons had so many large, bright barns that we would be able to film in. The key driver for me was the authenticity of the Nicholsons, how they look after their animals and how their animals are so important to them – not just as a commodity, but as living things."

"They are such genuine people that if you sliced them down the middle they're farming through and through," says Kim. "They're so proud and passionate about what they do and so willing to share their story. Even if the farm hadn't been so impressive – with television trickery it's always possible to make places look better – the one thing that you can't fake are the contributors. From the word go, Roger, Robert and David proved to be the real deal. When Roger talked about his love for his animals and what it means to be a farmer, he sent shivers down my spine."

Even though the farmers would already have welcomed the first wave of spring arrivals in early 2018, during the filming scheduled, around 250 sheep, 40 goats, 10 sows and several other animals were expecting to give birth, so it was going to be a very hectic time on the farm, even without factoring in a television crew, presenters and guests.

There's a showbiz saying that you should never work with children or animals, as there's always a chance of things not going quite to plan. With this in mind, several of the segments were filmed and edited in advance, ready to roll before the live programmes were made. Actor Kelvin Fletcher, who played Andy Sugden in the ITV soap *Emmerdale*, and former JLS pop star-turned-pig farmer JB Gill went to various farming locations around the country, filming segments about different aspects of rural life. Several reports about the challenges of farming in remote rural communities were also part of the line-up.

This being Yorkshire, it wouldn't have been right to have a programme without some familiar faces from *The Yorkshire Vet*, so vets Peter Wright and Julian Norton were also invited along to join in the fun. It was to prove very handy to have two vets on set, especially as during filming they luckily happened to have all their medical equipment with them.

When the programme was aired in spring 2018, programmes would go out "as live", which meant they could be recorded a couple of nights in advance, but when they appeared on screen it would look live as they would be filmed in one go. So there would be no stopping the camera for retakes if anyone fluffed their lines or anything went pear-shaped. Luckily the boys weren't thrown in at the deep end with a full length "as live" show; there was a lot of preparation work first.

"One of the very first reports we filmed was with Kelvin Fletcher, Robert and my dad," says David. "It was actually quite relaxed because there was just one camera operator there and a producer rather than loads of crew, so we all enjoyed it. Seeing Kelvin in action and his reactions to watching a lamb being born was a priceless moment." If you didn't see the first programme, it's worth tracking down the clip, as Kelvin goes from nearly vomiting at the sight of David's arm inside a ewe, to holding back the tears when he is allowed to cuddle a very sticky newly born lamb.

As well as the on-the-spot action, there were more structured film reports that would be spliced together for the live series. "When we first started and we had to record a film, we didn't realise that, if time allows, every camera shot is filmed three times," says David. "First as a close-up, then a wide shot, and then a listening shot, where you basically nod your head like a nodding dog. It was all a bit weird to begin with, but we soon got used to it."

So far, so straightforward, but the week the whole crew came to the farm was a different story, as Robert explains; "Having been fairly low key for all the videotaped reports, suddenly there was a massive crew of about 50 people at the farm, plus an audience of about 200 people watching us do our thing."

A few days before the first show went on air, engineers pitched up at the farm with a massive generator to power all of the television production equipment, plus a huge trailer which housed the editing equipment. Next the crew started to arrive – the director and producers, camera operators, set dresser, sound engineers, lighting engineers, floor managers, runners and make-up artists – with all their gear. "It all looked bloody expensive!" says Robert. "Which made us really want to up our game and not mess it up. I had to pinch myself that it was really happening to us."

Before they knew it, the day had come around for the very first show. But there was a problem. "We discovered there were glitches with the digital communication equipment, so more traditional equipment was set up. This meant there were electric cables strung all over the barn," says Robert.

"I had to hold a microphone, which looked something like Jimmy Hill would use, and commentate on David lambing a sheep. I felt a right chump."

As the programme was being filmed "as live", the brothers just had to keep calm and carry on. "It was all a bit surreal," says Robert. "I didn't even think about the fact we were going to be on television – having the studio audience in there with us was nerve wracking enough. They are listening and watching every move you make, so the pressure is really on."

Presenters Adam Henson and Lindsey Chapman had no sooner said "Welcome to *Springtime on the Farm!*" before Robert and his Jimmy Hill microphone were in action in the lambing barn. "An awful lot is happening here at the moment," he said. "There's a ewe here who's just delivered really lovely twins and we've got three other sheep in various stages of giving birth, so it's an exciting but chaotic scene at the moment. But we're hanging in here!"

It was a whirlwind of lambs' limbs as David cleared the airways of a newly born lamb, cleaned it with straw and placed it near its mum to bond; "That ewe licking the lamb gives it another 50 per cent in its chances of living. It's the elixir of life."

Later in the programme the celebrated television presenter Gloria Hunniford interviewed Roger and he recounted the story of Cannon Hall Farm, from the early years of struggling to make a living on the farm back in the old days to opening the tearoom and the decision to open the farm to the public. "We went through some dark, dark days," Roger recalled, "but since the farm opened we've never looked back."

Before filming *Springtime on the Farm*, Roger and the three brothers had never met Peter and Julian from *The Yorkshire Vet* – which is not really surprising as their veterinary practice is nearly 60 miles away on the other side of Yorkshire. But from the word go it was as if the guys had worked together for years.

"We had only just met Julian," says David, "and in the half an hour that we had to prepare for the show, he helped us with an emergency as one of our ewes had a prolapsed uterus. I had my whole arm in the ewe up to my shoulder and Julian saved the day. If he hadn't been there, we would have lost the ewe. The two of them are incredible and the more time that we spend with them, the more we learn from them."

Later in the second programme, it was the turn of Cannon Hall Farm's chef Tim Bilton to be interviewed by Gloria Hunniford, and later the two of them whipped up a batch of Cynthia's famous scones. Thankfully they used a tried and tested recipe so they turned out better than the ones earmarked for the Falklands all those years ago...

In Wednesday's episode, viewers saw farmers Roger, Robert and David turning out some of the new lambs into the field for the first time. "It makes me as proud as punch," said Roger. "It's such a joy to watch them and it makes all the late nights and bottle feeding all worth it when you see them running off into the grass. It's a great feeling for me."

On Thursday Roger was back in the spotlight, offering words of wisdom for wannabe farmers: "I wouldn't advise anyone to start off in hundreds of pounds of debt, that's for sure! Start in a small way, think of what you are going to produce, see if you can sell it and be prepared to give your whole life to it."

There were smiles all round at the end of the series with five days of jam-packed programmes completed. "It's been a week full of the joys of spring – quite literally," said presenter Adam Henson. As well as seeing new births galore, viewers got an insight into rearing rare breeds, hatching chicks, weaning pigs, shearing sheep, hand rearing goats and all manner of advice about farming.

As the crew packed up and said their goodbyes, Robert and David headed back to do more lambing. The television show was over, but in real life, springtime on the farm continued. "The day after the crew left everything felt very quiet on the farm," says Robert. "We almost felt bereft as we'd had so much adrenalin coursing through us for the whole week and then it all came to a stop."

It had certainly been a baptism of fire for the two newbie present-ers, having to tackle technical gremlins and know just what to say when the cameras zoomed in. "Rob did a lot of the talking," says David. "One day I totally hit a brick wall when I was being filmed and I just had nothing to say. I could see the cameraman was killing himself laughing at me and his shoulders were going up and down, but I just froze."

With hundreds of television channels that viewers can dip into, audience figures for a brand new programme are rarely off the charts to begin with, but there was a real buzz that *Springtime on the Farm* had been a success. People were talking about the programme on Cannon Hall Farm's social media channels and more and more visitors were coming through the gates.

With their newfound television fame came autograph and self-ie hunters. "When we first started being recognised it was really weird," says Robert, "but Yorkshire Vet Peter Wright gave us a lot of advice about going from doing a normal job, to suddenly being recognised from being on TV. He said to us 'You need to give each person as much time as it takes' which is sometimes hard when you've a million things to do, but the public is the reason why both the programme and the farm are so successful."

19

TV stars!

In November 2018, Daisybeck Studios got in touch with the fantastic news that, following the success of the first series, another run of *Springtime on the Farm* had been commissioned for Easter 2019.

As before, there would be lots of ewes ready to lamb, cows in calf and plenty of farrowing sows, but this time Robert and David had a plan up their sleeves that would make a fantastic addition to the opening programme of the new series. 2019 would mark 60 years of farming for Roger Nicholson, and his sons wanted to do something very special to celebrate the occasion. They were also going to buy some new breeds for the farm that were pregnant, so that, all being well, they would give birth when the programme was being filmed. Not only would it make the show more interesting because there'd be different animals for viewers to learn about, but it could potentially encourage more visitors.

Like the first series, the live action took place in the Roundhouse and the various other barns, and was again hosted by Adam Henson and Lindsey Chapman. Kelvin Fletcher and JB Gill had been invited back, and Peter and Julian from *The Yorkshire Vet* would also be joining the line-up.

It was another exciting series from the word go, as host Adam Henson explained in the first seconds of episode one: "There are literally thousands of farm animals across the UK giving birth right now, including in this lambing shed, where the ewe behind us is giving birth to a lamb as we speak."

It was the perfect start to the show. Next we saw Robert and David help deliver the 36th lamb of the day and then Roger's surprise 60th anniversary present was revealed. While Richard kept Roger busy back home in the farmhouse, Robert and David took delivery of a herd of shorthorn cattle – the same breed that had been on the family farm when Roger was a child. There was also a bull among the 18 heifers – the first bull to be on the farm for many years.

Reuniting Roger with his beloved shorthorns would be a fitting celebration for his lifetime of farming, but it was a tense moment when Roger first clapped eyes on his surprise present. If you think a bull in a china shop is dangerous, imagine what it can do on a busy farm. "Oh, my goodness me, what have you done?" Roger said as he saw the mighty bull nestling alongside the new herd of cows. But after the initial shock, he came around to the idea – not only having shorthorn cows back on the farm, but a bull, too. "I'm delighted to have them," Roger said smiling, "but there's been a lot of deception here. Was everyone in on the secret?"

It was the perfect way to celebrate Roger's career. "60 years has flown by and now the boys are doing a really good job taking it on, so I can't grumble. But now I have lots more cows to look after – so I can't retire either!"

In the second episode of the week, viewers got to see Robert in action at an auction, bidding on a Jacob ram to service the flock of Jacob ewes at the farm. Whenever he wanted to make a bid, he'd wink at the auctioneer. Later in the week viewers got to witness Roger's bidding technique at auction, as cameras followed Roger and Peter Wright on a road trip to Oban in Scotland to buy a Highland heifer in calf. "The idea is to be very non-committal, even if you're very keen on an animal," revealed Roger as they headed to the market." And Peter Wright had a good tip, too: "Just go in there as if you have short arms and long pockets!"

The two selected a beautiful Highland heifer called Fern that was in calf and would be a perfect addition to the farm. They knew that she would be pricey and Roger ended up paying a bit more than he had originally budgeted for. Catching the auctioneer's eye, he lifted his finger to say "100 more" and the hammer came down. Perhaps the spirit of his dad Charlie was with him when he too had offered that extra £100 needed to secure Cannon Hall Farm all those years before.

Back in the studio, viewers were introduced to the lovely Fern and her gorgeous fluffy calf Ted. "When I saw Fern, I knew there was something special about her," Roger explained. "When she walked into the ring, she had her head held high and looked like she was the queen of the ring."

Later in the programme, Roger put *The Archers* actor Charles Collingwood through his paces in the piggery, introducing a Hampshire boar to a sow. It didn't exactly go to plan, as the uninterested sows fled from randy Rocky the boar. Charles may be radio soap farming royalty as Brian Aldridge on *The Archers*, but he wasn't quite up to standard when he tried farming in real life.

Luckily, the rest of the series was plain sailing as Robert and David introduced us to new baby lambs, kids and calves. In the blink of an eye, another week of *Springtime at the Farm* was over and everyone could relax with a celebratory drink at the White Bull. *Emmerdale* actor Kelvin Fletcher certainly enjoyed the Yorkshire hospitality!

With the film crew dismantling their equipment, the props packed away for another time and the 'studio' turned back into a working barn, the team began talking about Series Three of *Springtime on the Farm*, not realising, of course, that things were going to be very different in 2020...

20

Coronavirus hits

On 16 March 2020, Government Health Secretary Matt Hancock told the House of Commons that all unnecessary social contact should cease. With the rising reports of outbreaks of Coronavirus, also known as COVID-19, it was clear that drastic action needed to be taken in order to fight the spread of the disease.

Rumours of the killer virus had already begun circulating months before, in a similar way that the SARS and Ebola diseases had come to people's attention in previous years. The World Health Organisation officially identified the outbreak of COVID-19 as a pandemic, but it wasn't until 23 March 2020 that Prime Minister Boris Johnson said people in the UK must stay at home and that certain businesses should close until further notice.

Having had to contend with the foot and mouth outbreak in 2001, the family knew how much shutting the farm was likely to impact the business. "We are involved with the National Farm Attractions Network and we took the lead with the decision to close early," says Robert. "I did a live broadcast to our Facebook followers saying that it doesn't feel right not to close. We could have remained open for a bit longer, but we thought it was better to be safe than sorry. I think we earned more in credibility and goodwill by shutting the farm."

It was an anxious time for everyone. Very early on, Robert and Julie's son Tom, who lives in London, had symptoms of the virus, but luckily no members of staff at Cannon Hall Farm were laid low with the disease. Roger and Cynthia had to shield at home in lockdown and other team members who were vulnerable immediately took the precaution of shielding themselves at home.

The longer it went on, the more worrying it became. "I was very anxious about the future," says Robert. "We had some savings in the bank, but you need money coming in to pay the wages and we had a hardcore team of full-time employees relying on us. I remember thinking, 'We are just turning on a tap here, paying wages with nothing coming in,' and I knew we couldn't last forever."

Fortunately, the Government announced they would be launching a furlough scheme, which could pay a proportion of earnings to take the strain off employers. For many businesses, including Cannon Hall Farm, it was invaluable. "The furlough scheme was a big weight off our minds. In the end, we didn't have to rely on it very much, but it was great for people who were shielding. Very sadly, one member of our staff, Nigel Elliot, became very sick, but being furloughed in lockdown had meant he could spend a couple of months of quality time with his family before he passed away. He was a really special member of our team, loved by humans and animals alike. He's irreplaceable."

As people tried to make sense of what was being called the New Normal, certain aspects of the farm continued as if nothing had happened. After all, animals still needed to be fed and kept in premium health.

In the farm shop, staff were busier than ever. "We were able to transfer staff over from the restaurant to the farm shop, take orders over the phone for collection and also make deliveries, so we were a lifeline for some people," says Robert. Turnover increased and the shop was so successful it received an award for the Safest Shopping Experience in the area. Yet another prize for the mantelpiece.

Obviously, the public wasn't allowed to visit the animals, but there was the small matter of the third series of *Springtime on the Farm* to think about. It was impossible for the camera crew to base themselves at Cannon Hall Farm, but with the magic of sophisticated mobile phones, the show could still go on. And behind the scenes, team members at Cannon Hall Farm were becoming quite the camera pros.

Having recently won two more awards to add to its collection, namely the UK's Best Large Farm Attraction and Best Digital Presence, Cannon Hall Farm was becoming a household name in Yorkshire, and with the success of *Springtime on the Farm*, its online following was going through the roof. In a matter of months, their already sizeable number of followers had doubled.

"We set up the farm's website in early 2000 and since then we've updated it as the farm has expanded," says Richard. "Then in 2011 an enterprise scheme in Barnsley helped us maximise our marketing potential on Facebook, putting us on the right track of how to tell our story. We spent a couple of days with an expert called Judith Hutchinson who showed us how to get the most out of social media and her advice has proved invaluable to us."

With Richard and Marketing Manager Nicola Hyde at the helm of the digital and marketing side of the business, the brothers decided to increase the number of live streaming broadcasts they were making for their social media followers, and also create a dedicated official Supporters' Group. "There were animals being born and all sorts of wonderful things happening and no one was there appreciating them," says Robert. "Therefore, increasing the live daily broadcasts was a way to keep in touch with our visitors and shop customers to let them know what was happening."

Dale Lavender, one of the farmers at Cannon Hall Farm, started filming and editing the videos for Facebook, and other members of staff also got involved so that there would always be plenty for farm fans to watch on their social media channels. And once Robert and the gang heard that lockdown would last as long as 12 weeks, they decided to commit to the daily live broadcasts so that everyone felt included in the day-to-day life on the farm. Whether that meant footage of hand-rearing lambs, or letting the shires have a run in the paddock, there was plenty to dip into.

"It was quite an undertaking making the commitment to film every day, but people seemed to love it," says Robert. "Almost immediately we started to get feedback from all over the world, which was amazing – and incredibly humbling."

With so much expertise in making films using just smartphones, it was decided that phone footage could be edited to a high enough standard that it could be shown on television. That meant that the third series of *Springtime on the Farm* would go ahead.

"Because Robert and Dave were so used to making films on their phones and Dale had such a good digital camera, they could set the camera up and communicate with the crew using a video call," Daisybeck's Kim Metcalf explains. "We'd then relay directions via bluetooth headsets. It was relatively simple, but it worked really effectively."

This time around, host Adam Henson was joined by former *Blue Peter* and Countryfile presenter *Helen Skelton*. But because of coronavirus and the demands of social distancing, Adam and Helen would be working from home, filming themselves outdoors and joining the Cannon Hall Farm gang by the magic of digital technology.

Robert and David reported on how they were coping with the pandemic at the farm: "We're incredibly lucky to have such a supportive online community here at Cannon Hall Farm – even from people who have never visited us. We've had such an outpouring of love and encouragement it's really lifted our spirits and made us feel more positive. We're going to look after the animals, we're going to do our best, keep the farm shop going and look after the business, and hopefully it will still be there when we come out the other side."

In the first episode, broadcast in June 2020 there was the tragic story of little Gordon the goat, who died 12 days after a very difficult birth. It was very sad to watch but highlighted the fact that for all the love and care that is put into farming, there will always be animals that don't make it. Happily, there was more positive news when Robert and David introduced two of the new pygmy goat kids, Dandelion and Burdock – the same names as the very first pygmy goats on the farm back in 1991.

Later in the week, viewers were introduced to Robert's nemesis, Zander, the feisty alpaca, and witnessed Roger's now famous 'Glove of Love' (see part two!) in action as he helped Lucky the boar finally get up close and personal with Bella the sow.

For the final show of the series, Robert and David showed off their carpentry skills as they unveiled Nannyland, a brand new adventure playground for the pygmy goats. "Farmers have always had to be multi-skilled because they're usually a one-man band and it's a tough job for a lot of people," said David. As he and Robert skilfully finished the goat's climbing frame David said, "I don't know about the A Team, Rob, I think we're the B team…"

And as a responsible goat owner, David did the honours in testing it out while Robert shook his head and said, "I think if those kids enjoy it half as much as you, we're in for some very happy goats."

In the final show, presenter Adam Henson summed up the series. "I have to say that despite all that's going on in the world, you two have managed to keep a smile on your face and bring us so much excitement for our springtime on the farm."

21

More TV tales

Watching beautiful lambs coming into the world, piles of the cutest little piglets suckling from their mum and pygmy goats trying out their new adventure playground was obviously just what the doctor ordered during the spring 2020 lockdown.

With viewing figures soaring and more and more people following Cannon Hall Farm on social media, there was a real buzz about the show. It seemed the *Springtime* team was ticking all the boxes for the kind of programme to really lift people's spirits during lockdown.

And even before the theme tune had played out on the last episode of series three, there were plans for a new television adventure. Channel 5 came up with the idea of *This Week on the Farm*, which would work in a similar way to *Springtime*, with the presenters reporting remotely. Again, footage at Cannon Hall Farm could be captured on smartphones.

This time, Helen Skelton anchored the show with Jules Hudson, who had previously worked with Robert and David in a programme about The Great Yorkshire Show. With a different kind of feel to *Springtime, This Week on the Farm*, broadcast in May 2020, saw the brothers reliving their childhood going camping, fishing and curlew spotting and even making gin. Viewers were also introduced to some of the most popular animals on the farm, including Shetland horse Jon Bon Pony and pygmy goats Charles and Camilla.

Coming up with names for the animals has been part of the enjoyment of welcoming new animals to the farm. Sometimes suggestions for names will be put out to social media, prompting brilliant ideas such as Pony Hadley and Pony Em. Other times, animals are christened in tribute to special friends and relatives. "It's one thing to call a Shetland pony something silly," says David. "But there's something much more noble about a shire horse. Calling it a jokey name just doesn't seem right somehow."

Presenters Jules and Helen both got namesakes in the form of Jules the donkey and Helen the alpaca. "It's was a huge compliment," says Jules. "And really goes to show that the four of us really do get on like a house on fire. The great thing about Jules the donkey is that whenever he appears when we're filming I can drop on one knee and he'll always come to me for a cuddle. He's such a warm, approachable fella."

Lockdown rules were finally relaxed in summer 2020 and in episode six of *This Week on the Farm* we saw presenters Jules Hudson and Helen Skelton reunited with the Nicholsons at Cannon Hall Farm. At that stage it obviously wasn't certain that all the drama of coronavirus was over, but the team was feeling more positive about the future and had adapted to the new way of working.

"In the past, for a lot of people, farming has been about that grumpy red-faced shotgun-wielding authoritarian who's going to throw you off his land," says Jules. "But with This Week on the Farm you get the fun of country life as well as some of the detail of farming."

Above: Charlie Nicholson loved to show his prize animals. **Below:** Sam the beer-drinking bull leads the winners

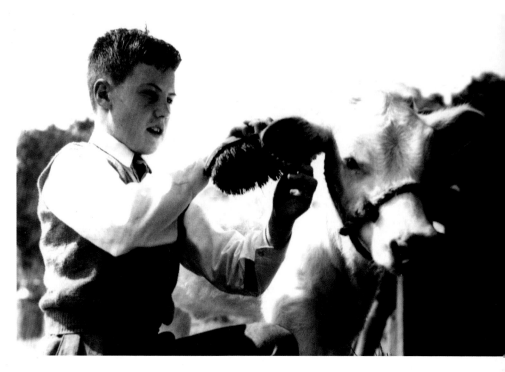

Above: In his Sunday best, Roger prepares one of the cows for a country show.
Below: Roger at home at Cannon Hall Farm, where he has lived since he was sixteen

Above: Richard, Robert and David playing on a woodpile. **Below:** The farm entrance, as it used to look

Above and right: Richard, Robert and David were ready to pitch in on the Cannon Hall Farm renovations

Below: Transforming the farm was a major project and at times there seemed no end in sight

Above: Jeremy the bull. **Below left:** Zander the alpaca. **Below right:** Dave and Rob feeding pygmy goats Primrose (white) and Millie

Stars of the show: Lucky the boar (above), Gary the donkey (left) and Awkward Orchid (below)

Right: Dave and Rob with Highland calf Ted

Below: Rob brushes down one of the show's popular characters, Ozzy Horseborn

A family affair: Roger and his three boys, Richard (left), David (top) and Robert

Once again, Channel 5 was delighted with the viewers' response to *This Week on the Farm*. And they weren't the only ones. In June 2020 Robert and David received a Points of Light Award and a very special letter from Prime Minister Boris Johnson congratulating them on their sterling work in front of the camera. It read: "Over the last few months, your daily broadcasts have allowed a nation in isolation to revel in the natural glory of Cannon Hall Farm, and to ramble alongside you in the spectacular Yorkshire countryside. Your ingenuity has fortified the nation's spirit as we continue our fight against Coronavirus. To do this while also running a working farm is wonderful and testament to the dedication of Britain's farmers, which you exemplify. On behalf of the whole country, thank you!"

Not surprisingly, the family were delighted with their Points of Light Award. "We said from the start of our broadcasts that if we could help one person, that would have been enough," Robert says. "We would have never expected that 13 weeks after we appeared on television we would be broadcasting live to thousands of people every day all across the world – and we've had so many lovely letters and emails from people saying how watching our broadcasts has helped them through this tough time."

David agreed: "We are so proud to have been able to share our family farm on a national scale – and represent our home town of Barnsley. But we do have to say it is a real team effort. We have lots of staff behind the scenes who have helped to make it a success as well. We are just the ones in front of the camera, but there's plenty of support behind the scenes so we will share this award with them."

2 2

Friday frolics

With lockdown over, but social distancing restrictions in place, the New Normal that everyone was talking about meant that visitors could come back to the farm, albeit in restricted numbers, but had to pre-book their tickets. Thanks to the success of the TV shows and the uplift in sales at the farm shop, Cannon Hall Farm was in good shape business-wise and the family looked forward to planning ahead for autumn and winter. From pumpkin picking to Christmas activities, the changing seasons have always been celebrated in style.

Members of the TV production team had talked to Robert and David about other programmes that could be based at Cannon Hall Farm – perhaps *Harvest on the Farm* or *Autumn on the Farm*, then they decided it should be called *Friday on the Farm*. In a similar way to *This Week on the Farm*, there would be pre-recorded reports from around the country, but the main part of the show would be filmed at night – and aired on a Friday rather than the usual Tuesday. It would have a more relaxed fun feel – with more of the Nicholson's now trademark awful jokes along the way...

Instead of the hustle and bustle of daily life on the farm, with visitors milling around in the background and the famers working on their day-to-day tasks, Robert, David, Jules and Helen revealed the magic of farm life at dusk. Mellow yellow fairy lights were strung around Roger and Cynthia's gorgeous garden to act as an outdoor studio, the indoor studio barn was softly lit with glowing lamps and the team donned warm clothes to keep them cosy in the chilly autumn evenings.

Once again, the programme would feel like it was a live show, but each episode was pre-recorded a couple of weeks before being transmitted. "This time we would be filming through the night, so it felt very different," says Robert.

"We filmed on either Sunday or Monday nights," says David. "Often we'd record the end of the show first when we still had plenty of energy, and the opening of the show last. In some programmes I think I look absolutely shot! I found it incredibly tiring, but it was a great thing to do and we wanted to show that just because the sun's gone down, there's still plenty going on at the farm."

It was an exciting concept, but had its risks, as Robert explains. "We knew that as the programme was going to be shown at 9pm that we would be going up against *Gardeners World*, which was BBC2's biggest programme and *Gogglebox*, which was Channel 4's. It was going to be a tough challenge to get good audience figures, but we like a challenge!"

On Friday 9th September 2020, presenter Helen Skelton kicked off the new series: "The nights are drawing in, it's a whole new farming season and we have a brand new time slot – what better way to kick off your weekend?"

And what better way to start a new series than with the big news many fans had been waiting for – Robert was getting a sheep dog. Usually brother David has the job of rounding up the animals, but

Robert joked that he wasn't always very obedient, prompting David to reply: "I'm looking forward to him not barking orders at me anymore!"

Pip the sheepdog was the ultimate star of the show and all the team welcomed the waggy new arrival, before it was time to catch up with Orchid the Shire horse and Helen the alpaca.

Then it was time for the first of the series' Dave v Rob challenges. The two brothers have always been competitive, so various activities were set up and threaded through the six-week run.

The boys were set a painting challenge which was judged by big brother Richard – who let's not forget has a degree in art, and he presented David with the winner's rosette for his watercolour of Fern the Highland cow.

Robert evened up the tally the following week when his Harvest bread showstopper won the next battle, but David fought back when he proved his worth at metal detecting by finding a silver sixpence. The challenge to create the best carved pumpkin was the order of the day for the Hallowe'en special, but as that was considered a draw, there needed to be some way to decide on a winner. And that's where the spooky Hallowe'en twist played out. The two brothers were sent into the woods to find pumpkin lanterns carved with their initials – but would they make it back alive?

It was a great end to another fantastic show – which also included a very painful team challenge testing the world's hottest chillies. David pulled the shortest straw and had to eat a 7 Pot Habanero, so called because one chilli is hot enough to flavour seven pans of sauce, but it was Jules who struggled the most, turning every colour of the rainbow as the chilli fired through him. "I'm an absolute wimp when it comes to anything hot," he says. "But I really really think there was a mix up and I got the hottest one! None of us could taste, speak or do anything for at least an hour afterwards! But it made great telly and that's the point. That's the great thing about Rob and Dave – they will pretty much try anything!"

Horse milk tasting was the order of the day the following week when the programme reported on the UK's only horse milking farm. Then it was back to Barnsley to test it against goats' and cows' milk. "It's sweeter and thinner than cows' milk and I really like it," says Robert. "Horse milk is considered the creme de la creme in certain cultures. In Azerbaijan there's a dairy that milks 2,000 horses and in Abu Dhabi there's a farm that milks 1,000 camels." No prizes for guessing what the boys would like their next taste test to be.

With an apple bobbing challenge which left Robert extremely soggy but victorious and a ploughing challenge which saw David taking home the gold medal, it was down to the final show of the series to even up the score. And what better way to celebrate autumn than with an old-fashioned conker fight.

First, find your weapon. Their passports may suggest Robert and David are adults, but the conker challenge brought out the big kid in both of them as they tried all their boyhood tricks to make their conkers unconquerable. Dave dipped his in vinegar while Robert coated his in yacht varnish to toughen up its exterior.

"I thought this isn't time for playing it safe," says Robert. "It's muck or nettles, go hard or go home. I've always had good hand-to-eye coordination, so I knew I had half a chance. And I absolutely smashed him!"

Although it was the last programme in the series, the story wasn't going to end there. "I've got a sneaky suspicion that there will be a few more Cannon Hall Farm competitions coming up in the not too distant future," said Helen Skelton, before revealing that a Christmas special was in the offing.

Bringing out all the big guns, the team behind *Christmas on the Farm* decided the programme would be a celebration of traditional festivities, mixed with a big helping of Nicholson nostalgia and Robert and David's now trademark competitiveness. Plus, of course, lots of animal action from many of the farm's favourite characters.

After such a testing year, the boys wanted to treat the farm to the best Christmas tree they could find – and what could be better than to buy it from the winner of the Best Christmas Tree in England? There was a wreath making competition – won by Robert – mince pie baking and a Victorian cooking celebration. And just in case there was any doubt that Robert and David were game for anything, the two of them dressed up for the Christmas card challenge, with a special appearance from Robert's granddaughter Nelly and a cameo role for Gary the donkey.

The donkey posse was joined by two tiny new friends, and Jeffrey and Prince the reindeer joined in the festivities and welcomed two young antlered chums. "Sadly we have to accept the fact that Jeffrey and Prince are old boys now," says Robert, "so it's good to have a couple of young ones waiting in the wings so they won't be lonely. There's one of each (a boy and a girl) so that we have the option one day of breeding our own little reindeer."

Even before the Christmas show was being filmed, Daisybeck Managing Director Paul Stead confirmed that another seven episodes of *This Week on the Farm* had been commissioned by

Channel Five, to be shown before the 2021 series of Springtime on the Farm. It was music to everyone's ears, as the team agree they have a winning formula.

"What's so nice about the lads is that TV hasn't changed them a bit," says Jules Hudson. "It's given them fantastic opportunities and allowed them to share their passions with a wider audience, but fundamentally they are just two lads who grew up on a farm in Yorkshire and that's what they love doing.

"Cannon Hall Farm is a really special place and Helen and I feel really blessed that we have a ringside seat and that we have been able to get to know the Nicholsons in the way that we have. They are just very lovely people."

"I have such an affinity with them both because their Famous Five childhood feels like mine," says Helen. "Just daft things like going fishing, silage times and playing with the cows. They are such warm, likeable people that it's easy to see how the nation has fallen in love with them."

Roger's reflections

The combination of decades of hard work to make the farm an award-winning tourist attraction and its small-screen fame has created an appetite for more television opportunities, interviews in the media and, of course, this book.

But it hasn't all been plain sailing and not everything has gone to plan.

Some aspects of the business have proved to be a difficult learning curve. There have also been sad goodbyes, health issues and difficult decisions to negotiate along the way.

Like any family, the Nicholsons have had emotional challenges as well as financial ones, but Roger's indomitable spirit continues to inspire the family to keep forging ahead. Here, he reflects on the highs and lows of his time at Cannon Hall Farm...

"I have a sadness about not being able to share my travel through life with my father and being able to ask him, 'Was I the real deal or did I let you down?'

"I was trying to be a successful farmer until we changed course and diversified and I now wonder if he would have approved of what we did, or whether we've gone too far with what we've done to the farm – all the concrete we've put down and the amount of new buildings that are now on what was once a lovely green field. Have we actually spoiled a lovely traditional farm?

"The thing is, I'm not sure there was an alternative to what we did because the farm just wasn't big enough to survive in a climate that was all about huge herds, huge turnover and amalgamated farms.

"I don't really think we had a lot of choice in diversifying. We had to do it if we were going to survive.

"Looking at it another way, I think my father would actually have appreciated what we have done and he was always such a friendly, amenable person. He would have loved meeting new people every day and showing them around the farm. In fact he could have probably done a much better job than me!

"I've lost two of my sisters, but I wish I could thank Olive, Shirley and Beryl for everything they did for me – in particular being so understanding about the farm coming solely to me when I was 21. In this day and age, inheriting a business just because you are the only male child in the family certainly isn't looked on as the thing to do, is it? But it was always going to be passed on to me and there was never a problem about it. My sisters knew that if we shared it out four ways between us there wouldn't have been a farm left, but there was never a cross word about that fact. They always supported me in everything that I chose to do throughout my life.

"Looking back, there have been some very difficult times – the first being when I lost my dad when I was 16. It changed my youth, there's no doubt about it, and I suppose I should be proud of being able to keep my head above water.

"The worst time for the family was definitely before we opened the farm to the public and the interest rates were sky high. At that point we didn't know whether we would ever be able to pay off our debts and embark on such an ambitious project. We got through by the skin of our teeth.

"Another time that was really challenging was in the 1960s, when it snowed so hard one winter and the sheep went over the walls of the farm and into the neighbour's garden. Normally they wouldn't have touched their poisonous rhododendrons, but because of the snow there was no greenery about. The following day I woke up to find 13 dead sheep. I also lost a cow and all of its calves to an incurable wasting disease called Johne's Disease, which was heartbreaking.

"2020 was very tough and took a bit of getting used to. Cynthia has been shielding and the business had to shut down for a while, so it wasn't a great time. We did our best just to keep going.

"Happy times far outweigh the bad times, though, and having the family growing up all around me has been the best thing of all. I'd like to think that we gave the children a nice place to grow up. We might not have had a lot of money to spend, but that's not always the be all and end all, is it? And as we've become more successful, life has become more enjoyable as it's taken the pressure off wondering if we're ever going to succeed.

"I'm not sure that any of the grandchildren will go into farming. Robert's eldest, Tom, works in London and his daughter Katie is the HR manager here at the farm. David's daughter Poppy is a make-up artist and Richard's son Marshall is still at school. He lives at our other farm in Barnsley and, like his dad, he's a great photographer. He always keeps an eye on the sheep for me though, so he's got the nickname Farmer Marshall.

"Cynthia and I have a great granddaughter now – Katie's daughter, Nelly Louise. I can't believe how good she is and it's absolutely fantastic spending time with her – she laughs all the time, she's always so happy. She likes the hens in the yard and the dogs, so I'm sure she'll be fine as she gets to meet more of the farm animals. Who knows, she may become a farmer herself one day..."

PART TWO

The Farm Today; Happy Days

S ince the 1980s, Cannon Hall Farm has steadily transformed from a traditional rural farm into one of the biggest tourist attractions in Yorkshire. But for all its daily visitors, social media followers and television viewers, it is, at heart, still a family farm.

The biggest change, of course, is that it is now a wonderful home to more than a thousand animals. From the smallest baby meerkat to the mighty shire horses out in the paddocks, the love and attention that the Nicholson family gives to their animals is a delight to behold.

Whether they are traditional farm animals like the 45 resident sows, three boars and 11 breeds of sheep, or there to enjoy their retirement, like Prince and Jeffrey the reindeer, all creatures great and small are cherished there. Some animals are bred to sell on to other farmers, while others end up in the farm shop; Cannon Hall Farm is both a traditional working farm and a visitor attraction.

The farm has diversified and thrived since owner Roger Nicholson first moved there back in 1959. There are now more than 250 full-time members of staff working across all the departments, be it the farming side, the restaurants, the sales and marketing team, butchers, bakers and candlestick makers (well, maybe not yet, but who knows?). The dedicated team at Cannon Hall Farm work hard to keep the success story alive, backed with support from thousands of loyal supporters.

"Obviously the farm has changed an awful lot over the years and there are always plans for new breeds of animals and new buildings for them to live in, so we're always evolving," says Roger. "The important thing for all of us is that we don't get too big, so that we lose track of what's important to us. It's a balance of both keeping a successful business going and not losing sight of what really makes us happy."

It would be impossible to mention each and every one of the animals that live at Cannon Hall Farm, but here is a selection of some of the favourites – and some characters you may not have met yet. We'll also introduce you to some of the many staff who work alongside the farm residents and reveal the magic that goes on behind the scenes at Yorkshire's favourite farm.

1

A is for Alpaca

" Z ander and I have a love-hate relationship," says Robert. "I love him, but he hates me."

If you're a fan of *Springtime on the Farm*, you will remember the priceless moment in 2020 when Robert first introduced Zander the alpaca to viewers. He and David had taken all of the alpacas out to the field to graze and Robert had put a bridle over Zander's head so that he could do a show-and-tell-session in front of the camera.

But when the time came to calmly walk Zander into the field for his 15 minutes of fame, the stubborn beauty had other ideas. "He's like a bloody bear," said Robert as the normally placid grey, black and white young camelid reared up on his back legs. Luckily, Robert had got used to thinking on his feet when unscripted incidents like this occurred and, as the restless fiend continued to wrestle with him, Robert quipped: "You know, in some civilisations alpacas are eaten for their meat – and I can fully understand why!"

That incident aside, Zander has become a firm favourite and is adored by everyone at the farm. That day he was just eager to join his alpaca friends who were grazing in the field. "He's a fine animal and normally we respect each other, but on that occasion I think he'd just had enough and wanted to get rid of me," Robert recalls.

The Nicholsons purchased Zander from the award-winning photographer Harvey Brown, who has been breeding and showing alpacas for several years. Harvey's farm in Colne in Lancashire appeared in the first series of *Springtime on the Farm* when former pop star JB Gill got a taste of competing in alpaca shows.

"As soon as we saw Zander we didn't want to come back with anyone else," says Robert. "He's the bee's knees and, at £3,500, he was also the most expensive alpaca we'd ever bought. He has such an outstanding colour and a look and presence about him. The moment we saw him, we knew he looked like an animal that people were going to fall in love with."

The name of the new alpaca was put to a public vote on the farm's Facebook page, though Robert and David already had a suggestion lined up for him. "We wanted to reinvigorate the name Dave as I think Daves are an endangered species!" says Robert. But the name Dave didn't get the thumbs up from fans. "The public went for Alexander, after Alex, one of the farm lads, and we decided to shorten it to Zander. It's a good, strong name, and we think it suits him."

Zander may be the most famous of the alpaca gang at Cannon Hall Farm, but he isn't the oldest. That honour goes to 29-year-old Barbara, who was born in Chile and came to England as a three-year-old. She is the mother of Adam, who was named after Adam Henson, one of the presenters of *Springtime on the Farm*. Barbara was well into her twenties when she gave birth to Adam and, as she was now too old to produce enough milk, the little cria (baby alpaca) had to be bottle-fed to survive.

Bottle feeding a baby alpaca sounds lovely, but it's not without its drawbacks. "I had to wear a coat when I went into her pen to feed

Adam," says David, "as Barbara would spit at me when I got close to her baby and her spit smelled like vomit. My washer did lots of extra loads because of her. But, of course, it was worth it to help little Adam."

Adam Henson's *Springtime on the Farm* co-presenter Helen Skelton also got a namesake when Audrey the alpaca gave birth. Like Adam the alpaca, tiny cria Helen also needed a helping hand to survive. "We nearly lost her three times," Robert explains. "I gave her a 10 per cent chance of survival because she wouldn't feed from her mum. We milked Audrey for four or five days to get the essential colostrum into Helen's belly, but she still went downhill very rapidly. It was staggering that we managed to bring her round at all."

After a trip to the vets to give her some glucose and a full check-over, little baby Helen returned to the farm for lots of TLC. "We become a little bit obsessed with animals that need extra help," says Robert. "We started feeding Helen every five hours with fresh goat's milk day and night, and slowly but surely she began to get stronger. Finally, thanks to the tenacity of the team, we were able to rear her to perfect health."

It takes around nine months for an alpaca to be properly weaned from milk to solid food, but because Helen had lived apart from the other alpacas she had to be reintroduced to them very slowly and carefully. "It felt bittersweet when I had to reunite her with the other alpacas," says Robert. "But we knew she needed to be part of a strong and happy community. I'd spent months hand-feeding her and I'd even spoken to her in alpaca to help her feel comfortable with me."

It was a big day when Helen was finally moved back into the alpaca and pygmy goat barn. "I was a bit worried that she would have an identity crisis," says Robert. "She'd been brought up by a Barnsley farmer, she'd been playing with goats and she'd never seen another alpaca since she was two days old. Oh 'eck."

Luckily, the reunion went totally to plan, with Helen's mum Audrey making a beeline for her chestnut-coloured daughter as soon as she saw her enter the barn.

The farmers stayed close to hand during the re-bonding sessions, in case there was any tension with the other alpacas. One in particular, Shakira, was initially a bit unwelcoming, but it wasn't long before Helen was learning to hold her own.

Alpacas are a relatively easy breed to look after – their main requirement is food and exercise. They look similar to llamas, but are shorter than them with blunt, rather than elongated, faces and stubby little ears. Their lovely soft fleece can be spun into yarn and shearing them in summer keeps them cool.

"Alpacas have become much more popular as a farm animal in the last few years," says Robert. "The main reason is their high-quality fibre. If you can get the very best black alpaca fibre to make cloth, all the suit-makers on Savile Row will pay through the nose for it."

In episode six of *This Week on the Farm*, Robert had a go at shearing 18-month-old Shakira – with some helpful tips from David. "We don't shear the tops of the head – one reason being that she's spitting like a good 'un at the moment! So we leave a pom-pom on her head and one on her tail." Shakira didn't exactly enjoy the experience, but the haircut was certainly something for Robert to be proud of. Even so, he won't be joining the shearing crew any time soon.

After her time in the spotlight, Shakira ended up looking just as much of a rock star as her celebrity namesake. But not all of the Nicholsons' alpaca collection were always quite so glamorous.

Audrey, for instance, had two parallel sets of teeth as her baby set had never fallen out. "She certainly wasn't named after Audrey Hepburn, that's for sure!" says David.

Fortunately, it was a painless operation for her baby teeth to be cut off with a cheese wire, after which her long adult teeth were filed down to give her a Hollywood smile. It sounds painful, but alpacas have no nerves in their teeth and feel no pain while the drastic looking dentistry is taking place.

Zander also had to have his teeth filed, which was shown in an episode of *Springtime on the Farm*. Thankfully, he behaved better than he did on his television debut – the fact that he had a foot-long file in his mouth may have had something to do with it.

As with many of the new breeds that the Nicholsons have welcomed to Cannon Hall Farm, the farmers have learned more about alpacas through experience. And as the farm has become more and more successful, the family has been able to upgrade their collection with better stock.

"When we bought Zander, the breeder gave us great advice on how we could improve the quality of our herd," Robert explains. "The best tip he gave us was to buy a really good stud male – the best we could afford – so that every baby that is born will be an improvement as the bloodline gets stronger."

Alpacas breed all year round and are originally from the high plains of South America (hence Shakira's Latina name). They communicate through body language – clearly demonstrated with Zander showing Robert who was boss when he reared up on his back legs. They spit when they are stressed but make a delightful murmuring noise to get attention.

Alfie the alpaca gets the prize for being the friendliest of the bunch at Cannon Hall Farm. Robert and David were doing their morning rounds out in the fields one day in June 2019 when they found a beautiful pure white newborn cria trying to feed from Gary the donkey. It was a stunning summer's morning and Gary and the donkey gang didn't seem the slightest bit perturbed by the little visitor who had come over to say hello. Equally, the cria wasn't at all nervous as David scooped him into his arms and reunited him with his mum, who was sharing the field with the other alpacas and donkeys.

Back at the farm for a quick health check and some one-on-one time with his mother, the little alpaca immediately bonded with everyone around him – including young farmer Tom. "It's not often that you get an alpaca as friendly as Alfie, so he's one of a kind. He's definitely my favourite – he's got such a character – like no other animal," says Farmer Tom.

In September 2020, beautiful alpaca Beyoncé gave birth to a tiny little girl cria, but she was a cause for concern from the word go. Apart from being very small, the cria was incredibly weak. Shona the vet was called and she wanted to get the cria to the surgery to investigate further, but the little alpaca needed strengthening up first. She administered steroids to give her immunities a boost and instructed Robert and David to milk her mum Beyoncé for the life-giving colostrum the baby so urgently needed.

When Beyoncé failed to produce any milk for her baby, it was a sure sign that the cria was premature. A mixture of sheep and goat's milk was then fed to the newborn and for a while things seemed to be more positive.

Robert and David fed the cria at two-hourly intervals and David seemed hopeful that she was getting her strength: "I really thought that she had a fighting chance. But the next time we checked her, she had died."

"I think that makes it worse," says Robert. "With any newborn, once you feel that you're in with a chance, it's a much bigger blow when they don't make it. If it's a lost cause from the start, you somehow get used to it, but some fights you just can't win."

But for all the sadness of losing Beyoncé's cria, there was soon a new addition to the alpaca gang when Whitney went into labour with her new baby alpaca, Chino. "She gave us a real boost just when we needed it, but that wasn't without its difficulties either," says Robert. "She had a shoulder that was fused together and I had to give her physio to get it moving and bottle-feed her at first. But once she'd had that first vital feed she was on her way."

David adds: "When the alpacas lose a cria we are often more upset than the cria's mother is, because they are such resilient animals. But that's the nature of farming. As hard as we try, there will always be animals that don't make it, but every new life on the farm is so special to us."

At the end of 2020, the farmers said goodbye to three of the alpaca gang in an episode of *The Yorkshire Vet*. Sammy, Dean and Frank, better known as the Bachelor (al)Pack, went off with llamas Bella and Pedro to become therapy animals in Northern Ireland. "It's a different kind of sad goodbye," says Robert. "But we know they'll bring joy to all those they meet."

2

The Anniversary Herd

The Shorthorn cow may not be as instantly recognisable as other breeds of cattle, such as the black and white patterned Friesian or the hairy Highland, but it has a very special place in Roger Nicholson's heart. The breed originated in the North East of England in the late 18th century and was developed as a dual-purpose animal, meaning it is suitable for producing both milk and beef. For this reason, Shorthorns have always been useful additions to any farm, and they were a big part of Roger's childhood when his father Charlie was at the helm at the family's farm in Worsbrough Dale.

One particular Shorthorn, known as Sam the beer-drinking bull, was Charlie Nicholson's pride and joy, winning various country shows throughout the 1950s. He was a magnificent example of the breed and his legacy lives on at Cannon Hall Farm, as the White Bull restaurant is named after him. Nevertheless, for several years, Shorthorns weren't part of the farm's animal collection. The family had tried various other ways to make the farm viable, keeping dairy cows at one stage, then breeding pigs, but Shorthorns hadn't been part of the equation.

In 2019, the three Nicholson brothers decided to celebrate

Roger's 60 years at Cannon Hall Farm with a very special present – a brand new herd of Shorthorn cattle. "We wanted to mark the occasion with something very special," says Robert, "and we knew Shorthorns were his favourites. By chance, a friend of mine had some that I could buy from him and they were first-class stock, so we struck up a deal.

"We hadn't had a bull on the farm for about 40 years, as we never had enough breeding stock to justify keeping one. When you only have a few cows, it's too expensive to keep a bull full time, so artificial insemination is used instead."

This time, though, there were going to be 18 heifers to be serviced and so the herd included a hefty handsome bull called Jeremy. It was hoped that, all being well, the new arrivals would breed plenty more pedigree cattle that could be sold on.

The first programme in the second series of *Springtime on the Farm* featured an elaborate ruse that the brothers had to pull off in order to keep Roger busy while the new herd arrived at the farm. On the day of the big reveal to Roger, Robert and David even managed to tie a pretty blue bow on Jeremy – not that there would be any doubt he was a bull... Meanwhile, eldest brother Richard pretended to record a social media interview with Roger back home in the farmhouse until the cattle had been delivered and lined up ready for inspection.

All the brothers waited anxiously to see their dad's reaction when he was allowed to see what the boys had cooked up. David certainly had his reservations. "The worst thing that could happen is if my dad just doesn't want a bull on the farm. Bulls can be very dangerous, and my dad just might not want one."

But it was too late to change their mind. "We're in for a penny, in for a pound," said Robert. "They're ordered and they're on their way. We just have to hope that Dad reacts in a positive way."

As Roger came into view, shielding his eyes from his surprise, everyone held their breath. When he was allowed to look, it took him a few seconds to take it all in. Happily, Roger was thrilled with his anniversary gift. "Thank you everybody for this kind thought," he said, once he'd got over the shock. "I shall really enjoy going into the cattle shed at night to see my new cows. Just them and me."

The family looked forward to seeing the herd increase – but all in good time. "We decided we didn't want all of them to be in calf at the same time, so we kept about half of them in a different field, so that Jeremy wouldn't impregnate all of them straight away," says Robert. It was then just a matter of letting nature take its course.

During filming for *Springtime on the Farm*, vet Julian Norton scanned the heifers to see if they were in calf, and everyone was delighted when he confirmed eight out of nine of them were.

Unfortunately, though, not everything went to plan. Soon afterwards, one of the heifers started calving seven weeks early. Instinctively, David thought her calf had died inside her and sadly a vet confirmed this to be the case. Her stillborn baby calf had to be removed by Caesarian. And there was more bad news to come when a second heifer also went into labour early and lost her calf. "They both had a high temperature and we think the two of them must have picked up some kind of virus which caused the loss of the two calves," David explains.

As the mothers who had lost their calves were still producing milk, it was decided to foster a newborn calf on to each of them. "The calves we use for this purpose are usually from a dairy farm, but one of them came from one of our own cows that didn't have enough milk to feed its calf," Robert says. By carefully, gently introducing them to each other and supervising them, in case there's a negative reaction from the heifers, it seemed to work. "For us, it was making the best out of a bad situation."

When the third of the new heifers went into calf early, the brothers were concerned that it might result in another stillbirth. Would this mean that the whole herd had been infected by a mystery virus? "To lose three calves would be a disaster," said Robert at the time.

To make sure everything went smoothly, Robert called in David the vet to help the struggling mum. A winch known as a calving aid was used to help move the birth along and the young calf entered the world in perfect health.

"We had the full range of emotions with the first few births," says Robert. "It was such despair at the start because we were thinking, 'We're just trying to do something nice here', and everything seemed to be going wrong for us. We'd thought they were an easy breed to look after, but they were anything but."

When the next heifer went into calf, vet Matt and farmer Ruth were on hand to help deliver a healthy calf – a lovely little white bull. Unfortunately, during the problematic delivery, its wrist was broken. "Breaks can happen, despite everyone's best efforts," says Robert. "We were concerned that once it was mended the leg would never be strong enough for the bull to be able to serve the females, as it wouldn't have the strength to jump on their backs. But we were highly delighted as he made a really good recovery."

"He looked a good strong calf and that's what we'd been waiting for," says Roger. "It's a pity that we had that little complication, but when they get trapped in that position they don't last very long if they're not pulled out. Once we knew that he was healthy and that he would survive it was such a great feeling to have a white bull back at the place."

Roger was as pleased as punch to welcome the little white bull, so it seemed fitting that the new addition should also be called Roger. And there were more reasons to celebrate when Roger the bull's mother went on to have another bull calf. This one needed no intervention from any of the farmers or the vets. In fact, it was such a good specimen of the traditional red and white colour of the Shorthorn, that Roger, Robert and David earmarked him as a breeding bull. Good news financially and for the breed itself.

As no more than one breeding bull is needed at a farm, Roger and the other magnificent male Shorthorns were sold on as pedigree breeding stock. "Roger now belongs to one of our neighbours and we see him out in the fields on a regular basis," says Robert. "He looks an absolute belter."

The last of the first batch of anniversary heifers that Jeremy made pregnant gave birth out of the blue one day, and farmers Darrell and Ruth were on hand to make sure all was okay. "She did it all by herself," says Darrell, "and the calf was up and suckling and being cleaned by her mum in a matter of minutes."

Fortunately, when it came time for the second batch of Jeremy's heifers to be delivered, everything went off without a hitch. "We've learned that buying a herd of heifers is lovely on the one hand, as we will have them for several years to come, but first-time mums often have problems," says Robert. "Happily though, since Roger the bull, we've had 12 new calves and they have all been fit and healthy. And all bar one has been a natural birth, which has been such a happy ending to the anniversary herd story. It's been exactly what we hoped for when we bought them for Dad and it's been nothing but a pleasure, even though the first year felt like nothing but difficulty."

After Jeremy's Summer of Love in 2020, Robert and David were confident that their second batch of heifers were all in calf. "Jeremy is in the prime of his life – about five or six years old and he's doing a fantastic job for us and will be doing so for some years to come. He's an energetic bull and, although he's not young, there's life in the old dog yet."

3

Mary from the dairy

lthough Cannon Hall farm isn't technically a dairy farm, there's a daily milking demonstration for visitors in the milking parlour. The star of the daily demos is Mary from the Dairy, daughter of the original farm dairy cow, Elsie.

Mary, who produces around 20 litres of fabulous creamy milk a day, is a typical Jersey cow, with beautiful big oval eyes, outlined with darker circles than her lovely caramel coloured coat. "When we opened the farm to the public, we decided to do milking demonstrations because it's fascinating to watch if children are unaware of how milk is produced," says Robert. "We figured that we would have different types of cows and talk about their various attributes, then in 2013 Mary was born, who not only looks wonderful but, as a Jersey cow, makes a high butterfat milk which is great for cream production and butter."

As Cannon Hall Farm is not a commercial dairy farm, Mary keeps her calves with her for longer than traditional milking cattle, as she produces enough milk to feed both her calf and provide milk for the milking demonstration.

In 2017, Mary developed a mastitis infection in one of her teats and, as hard as the farmers tried, they couldn't save the teat.

Thankfully, though, they were able to save the other three. "She's what old-fashioned farmers call a three-papped cow," Robert explains. "Incredibly, her mother Elsie had exactly the same condition, so maybe it's in the genes. She's always managed to produce a great amount of good quality milk, though."

Mary has had more than her fair share of dramas over the years, but she's a strong female and is obviously very happy at Cannon Hall Farm – even if she sometimes gets a little confused. "One day I found Mary fussing over what I thought was her newborn calf, but the calf actually belonged to a Shorthorn cow," says Robert. "Because Mary was so close to calving herself, her maternal instincts had kicked in big time and she'd stolen the calf and was licking her and bonding with it."

While the real mum looked on rather bemused, Robert and David were concerned that if Mary continued to lick the young calf, its real mum would reject it. "On top of that," says David, "when Mary's calf was born, there would be no end of confusion. We had to work quickly to get the calf away from Mary, without stressing out either of the cows who were getting very upset by this point."

While David picked up the calf and encouraged its real mum to follow him, Robert had to keep Mary at bay while the poor cow looked on wondering why 'her' calf had been kidnapped. It was hoped that Mary would soon settle, but she became extremely stressed and wouldn't calm down. Worried that she may damage her unborn calf, the farmers moved her to the pen next door to the Shorthorn mum and calf, hoping the proximity would at least bring her a little comfort.

"The next day when we went to check on her, we found that overnight Mary had had a little bull calf and mother and baby were doing very well." David says. "It was the perfect end to the story. And with Mary's track record of producing the very best milk, her calf was about to have the very best start in life."

4

Donkey drama

From the word go it's been a rollercoaster of emotions for the donkey dynasty at Cannon Hall Farm – mainly due to a certain little character called Gary.

With his beautiful almond-shaped eyes, sprightly ears and lovely soft coat, he is the epitome of the perfect donkey, but he was a little devil in disguise at the start of his life at the farm.

Robert and David adopted Gary from a travelling community in Doncaster in 2015, and things were tricky from the word go. "You could say Eeyore-ways was a problem," jokes David.

Cannon Hall Farm already had a number of large French donkeys, but their little English donkey Derek had never proved the best mating partner, and Robert and David reckoned a big stallion like Gary would be just the job. "We had to pay £600 for him, which was quite a lot, considering he was so hard to deal with and so awkward," Robert says. "He was very hard to like, let alone fall in love with."

The farmers don't know what kind of life Gary had in his early years, but he certainly hadn't had any training to tame him in any way. "He was very stubborn and it took us a long while to settle him in," Robert says. "The first time we took him to the farrier was a nightmare. He was really uptight and we had to sedate him in order to be able to trim his hooves. He really didn't like that kind of close contact with people."

Donkeys' hooves are much more likely to absorb water than horses' hooves, so they thrive in warm dry climates. In soggy old England, meanwhile, they need a bit of extra help. Spending time on lush soft pasture means their feet are more likely to suffer from foot diseases, and foot trimming is normally needed around every couple of months to cut back the absorbent areas of the hooves and keep the area dry.

With a certain amount of donkey whispering and careful training from equine expert, farmer Ruth, back at the farm, Gary started winning friends – and soon showed that he definitely had a way with the jennies (female donkeys). As soon as he was unloaded from the horse box at Cannon Hall Farm, he set to work with the donkey ladies. He noisily announced his presence and dispensed with the niceties as nature took over. It would only be a matter of time before they would be hearing the gentle pitter-patter of donkey hooves.

As with all animal births at Cannon Hall Farm, the farmers try not to intervene in any labours, unless there are signs that something is going wrong, which is what happened when beautiful donkey jenny Daffodil was giving birth. "I'd never assisted a donkey birth before," David remembers. "For a while, farmer Ruth, farmer Darrell and I just watched on the sidelines to check everything was going okay. But the longer it went on, the more we could see that Daffodil was really struggling and that we needed to help her."

David gently tried to ease out the foal and made sure the front two feet and the head were aligned ready for birth. Then, with some force, he pulled at the newborn's hooves, while Ruth gently stroked

Daffodil's head and tried to soothe her. It was obvious that the foal was going to be huge and every contraction seemed to cause more and more distress to the first-time mum.

Finally, with help from farmers Ruth and Darrell, the foal was heaved out of its mother and Daffodil immediately started licking her beautiful young foal, Doris, clean, naturally bonding with her. "It took us quite a while to help Daffodil and it was a really, strenuous effort," David says. "The baby was well and truly stuck and it certainly took some finessing to get it out safely."

When Gary's next two foals came into the world, vet Julian Norton came over to the farm to check them over. Julian was very taken by them, so it seemed only fair that one of them should be named after him. The other one, a baby jenny, was named Juliet.

Juliet's mother, Deirdre, was quite an old girl. Her breeding days should have been over by that stage, but feisty Gary was continuing to bother her, so the farmers made the decision to move her and Juliet on somewhere quieter. "We found a fabulous place for them in Hampshire," says Robert. "The lady who adopted Deirdre and Juliet already had two donkeys of her own and one of them was 35. She wanted another donkey to care for when the inevitable happened and the older one went to the donkey sanctuary in the sky."

Donkey Julian, meanwhile, found his new forever home in Halifax. "You often find that donkeys work very well with people who need a bit of love and companionship in their lives, and Julian went to a lady who had some troubles with anxiety." It was a lovely happy ending.

Unfortunately, not all donkey stories at Cannon Hall Farm have been quite so straightforward and it wasn't too long before Julian the vet was getting hands-on with Gary. In a now infamous episode of *The Yorkshire Vet* – which in turn was replayed in eye-watering detail on Channel 4's *Gogglebox* – Julian was called in to examine Gary's penis. The donkey had a nasty growth that was infected, ulcerated and bleeding – but in order to get a better look at the problem privates, Julian needed to see the full, erm… extent of the problem.

One of Gary's girlfriends was brought into the pen to see if it might encourage him to show more of his nasty nethers – but things didn't quite go to plan and the vet nearly got a hoof in the face for getting too close. Julian then sedated Gary so he could get a better look at the extent of the problem penis. "Rubbing a donkey's penis was not quite what I had in mind when I was having my breakfast this morning," he said as he injected the delicate area and massaged the anaesthetic around.

"Julian then decided to do some old-fashioned veterinary care and stretched some rubber bands around the top of the growth – the ones we put on lambs' tails to make them drop off," Robert explains. "He then let the magic of nature take over, hoping that the growth would drop off and all would be well."

Julian suspected Gary had a sarcoid tumour, which is quite a common condition in equines. Although the word tumour suggests cancer, sarcoids are classed as low-grade tumours and don't spread internally.

Six weeks later and hey presto, the infected area did indeed drop off. Gary's nether region healed up and all was well – until six months later, when the infection came back with a vengeance. "It came back in a much bigger way and so we got Julian the vet back to look at it and he and our local vet Matt operated on him."

This time, crossing all fingers, it looked as if Gary was okay and he went on serving the jennies. Unfortunately, it then happened again – the swelling was back. "At this point we made the decision that Gary couldn't keep having major surgery as it was possible it would keep reoccurring," Robert explains. So the old rubber-band method was used again, and would be on subsequent times if it was to ever to come back again.

With the rubber band trick seemingly working, Gary was back to serving the jennies and soon a number of them were pregnant again.

Gary's tackle was obviously tip-top, as Crocus gave birth to a whopping foal in July 2020.

Also in July 2020, Daffodil the donkey went into labour again. "Donkeys usually give birth late at night, but in the early hours of the morning we went to help Daffodil as she'd been pushing for a long time," says David. "From start to finish, an equine birth usually only takes around half an hour, but it was obvious that this foal was really big and was causing major grief for its mum."

Finally, after around another 20 minutes, the foal was freed and everyone breathed a sigh of relief as Robert and David checked over the beautiful male baby foal. He was an absolute sweetheart with enormous Bambi eyes and a golden fluffy coat. Given that the brothers had got to know vet Peter Wright so well during filming, they decided the new donkey should be named after him. "Daffodil and the foal should have been absolutely fine," says David. "But for some reason Daffodil wouldn't bond or try and feed Peter. In fact, she was really hostile and she tried to kick him and bite him."

It was heart-breaking to watch as Robert and David persevered with Peter, putting him into a pen near Daffodil, hoping that she would learn to love him. It was also important that the little foal drank as much of the mother's colostrum as he could so that he would be able to fight infection. Luckily, he was a strong suckler and took every advantage he could to feed from his mum, so the farmers hoped all would be well.

Unfortunately, though, things got worse, as Daffodil really didn't want to bond with her foal and kicked out at the little newborn. Poor little Peter looked so defenceless as Daffodil rejected the foal, and Robert and David made the decision to move him out of harm's way and hand-feed him.

Peter was weighed in order to work out how much food he needed. "We realised why it was such a difficult birth for Daffodil because at more than 22 kilos, he was such a significant size," says Robert. "He would need 10 per cent of his body weight in milk every day to survive."

The brothers hoped that Peter would have suckled enough colostrum to give him a good chance at survival. He fed well the following morning, but by the time farmer Ruth went to check on him at lunchtime, things had gone downhill. One of Peter's ears was drooping and he had a high temperature. On the vet's advice, David quickly took him to the animal hospital in Huddersfield. Everyone had hoped that Peter would pull through once he was given the nourishment he needed to survive, but two hours later, the boys received a heart-breaking call: tragically, little Peter had passed away. "It had never entered our heads that Peter wasn't going to make it," says Robert, "and we'd grown attached to him so quickly, he was such a bonny little thing."

Vet Peter Wright said the donkey's demise could have been due to a number of issues. "Peter was a particularly large foal and probably quite painful to produce, and as Daffodil looked at him she probably thought that's the result of my pain and that's possibly why she rejected him," he said. "The fact that Peter didn't get sufficient colostrum and possibly the milk quality wasn't as good as it needed to be sadly resulted in Peter getting an infection – most likely septicaemia, which Rob and Dave could do nothing about."

It certainly made tough viewing when the story was shown on *This Week on the Farm*, but once again it underlined the fact that in farming nothing is ever guaranteed. "We had two wonderful donkey births in 2020, with Jules and little ginger-eared Kim, but sadly we've lost two other donkeys as well," says David. "Of course, we'd like them all to live and although we delivered them all well, there are a number of factors why they didn't make it. It was devastating for us at the time, but we now have six healthy breeding females and they are all pregnant by Gary. So summer 2021 will be a busy time for us."

In the 2020 *On the Farm* Christmas special, the donkey crew was joined by six beautiful new miniature donkeys from Northern Ireland – three jennies and their foals. "They were quite expensive, as they are very sought after, but we thought they would make a fabulous addition to the farm," Robert explains. "They're only 32 inches to the shoulder, so they really are tiny, but like our Shetland ponies Jon Bon and Ozzie have proven, good things come in small packages."

5

Prince and Jeffrey the reindeer

There are surprises at every corner at Cannon Hall Farm, with several breeds of animals you would never expect to see in the middle of Yorkshire. Two such characters are the resident reindeer Prince and Jeffrey, who are enjoying their retirement in their specially designed spacious paddock.

The farmers bought the two old boys from their previous home in Essex, where they had enjoyed a glittering career of movie and television work, not to mention seasonal guest appearances accompanied by a rotund white bearded chap in a fetching red suit. If you are a particularly good reindeer spotter, you may have seen Prince, the pure white reindeer, on advertising hoardings around the country at Christmas time advertising various top-notch department stores.

But with their careers behind them and now needing a less taxing lifestyle, Robert and David were keen to get Prince and Jeffrey to Yorkshire. "We decided that their days of pulling Santa's sleigh were over," says Robert, "and we knew that our visitors would fall in love with them."

Prince and Jeffrey arrived at Cannon Hall Farm in October 2019 and Roger was particularly smitten. "They were a little bit wary of me at first," he says. "But they are easily seduced with food."

Reindeer in the wild graze on lichen, moss and grasses, and this is replicated in special reindeer pellets, supplemented with soaked sugar beet pulp and plenty of carrots (whole ones for Prince, chopped chunks for toothless Jeffrey). "I've really taken to them and it's mostly my job to look after them," says Roger. "They need a bit more care and attention than sheep and cattle, but we give them exactly what they want."

Yorkshire winters can be very harsh, with temperatures often dropping below zero degrees, but with their fluffy nostrils to warm the air they breathe and furry feet, reindeer are well equipped to cope as temperatures can get as low as -30 degrees in the Arctic.

At Cannon Hall Farm, Prince and Jeffrey are kept in a limestone paddock rather than let out grazing in the fields, as they are susceptible to worm infestations.

They also have a night shelter – which came in particularly handy in February 2020. Prince's right eye had been looking very swollen and sore, and vet Peter Wright, along with Matt from Donaldson's Vets, agreed that surgery was needed. The swelling was an ulcerated tumour and to give Prince the best chance of survival, it urgently needed to be removed, along with the infected eye. It was a major operation, but unless the whole eye was removed, the cancer was very likely to spread.

"Peter explained to us that the success or failure with Prince's treatment was not so much reliant on the operation," Robert says. "It was down to how Prince came out of the anaesthetic. "You have to keep him warm and Peter had brought a duvet along with him to wrap up Prince after his surgery. Dad and I also rigged up a heat lamp in the reindeer' night shelter."

Taking to social media that night, Robert explained how nervous he was about Prince as he filmed the motionless antlered patient. A few hours later, at two o'clock in the morning, Robert visited him again and was overjoyed to see Prince up and about. He was looking a bit confused, but he was strong and very much alive. "It was a real tribute to Peter that his old-school technique of bringing along a duvet to keep Prince warm after his operation was definitely what saved him."

A month after the surgery, vet Matt returned to remove the stitches from Prince's eye and gave him the all-clear. "We were so delighted that Prince healed so quickly," says Robert. "Going through the drama of his surgery has made him even more special to us. We're always careful not to approach him from his blindside as that tends to make him jump, but apart from that he is in tip-top health."

The two elder gentlemen reindeer look very serene as they watch the world go by from their dry paddock at the farm, looking out over the lush south Yorkshire fields. When visitors to the farm see them, they are often surprised to see reindeer this side of the North Pole, but the two handsome stags take it all in their stride. "Jeffrey has always been the bossiest of the two of them and continues to be top deer," says Robert. "But they are great companions and now, like so many of our animals at Cannon Hall, our reindeer are real farm favourites."

6

Victor and the legendary llamas

Visit Cannon Hall Farm and you are sure to see a photograph of Victor the llama chomping away at some hay. The much-loved creature was the first llama to become part of the Cannon Hall family when the farm opened to the public.

As much as people enjoyed coming to see him, he didn't always share the love… "One spring we had a local ITV news team at the farm," David recalls. "They came to film various things going on and decided to take a closer look at Victor. The presenter, Marylyn Webb, was interviewing us and she was wearing a lovely expensive coat when Victor decided to put his head back and spat horribleness over her lapel. It was so embarrassing. He didn't spit at many people and I don't think he really took against her – I just think he didn't like the look of the fluffy microphone hanging over him."

Every year, Robert and David have to give the llamas worming treatment if they are out in the fields all summer. As each of the fully-grown herd weighs in at around 30 stone, the boys have to carefully get in close among the herd – within spitting distance in fact…

Like alpaca spit, llama spit is a foul mixture of saliva and gastric juice. When a llama pulls its ears back flat against its head, stares intently, raises its chin and starts to gurgle, it's the sign to duck out of the way – quickly. Spitting is a way of establishing the pecking order and is a recognised feature of llamas, which are one of the most domesticated animals on the planet. Llamas have been working with man for around three thousand years and are closely related to camels and alpacas. They are all classed as "even-toed ungulates" and also share many of the same characteristics as wild guanacos.

"After the first Victor, we got a new collection of llamas from an exotic animal breeder," says Robert. "When we bought them, I remember thinking they were quite cheap – I had been expecting to pay five times as much for each one. Then, about six months later, we found out why – we'd bought guanacos, not knowing the difference between them and llamas."

Llamas are much taller and more domesticated than guanacos, which are native to arid regions of South America. Llamas can also be various colours, whereas guanacos tend to be a dull brown with a paler underbelly. It's not that the Cannon Hall farmers had anything against guanacos, they just thought that llamas were more interesting to look at. In time the guanacos were upgraded to their more handsome cousins.

A short while after the first Victor died of old age, Victor number two was welcomed to the farm, followed by Elvis, Priscilla, Dolly, Dusty and Busty (named by David because, he says, "she's a bit of a handful!").

Priscilla's cria Spindles, known to television viewers as Lisa-Marie, had a difficult start when she was born out in the field in the pouring rain. The newborn was spotted by a member of the public, who alerted the brothers to rescue her. "It was a moderate emergency," says Robert. "We had to fetch her in, dry her out, get her fed and try to get her to bond with her mum. We then had to bottle-feed her three times a day as her mum was a little bit short of milk. But with a top-up of goat's milk, she was soon thriving. She looked the image of her father, Elvis."

Most llamas have natural births. "They sometimes need a bit of a helping hand," says Robert, "but they are generally quite streamlined and slip out quite easily." Such was the case in 2008 when young cria Kayode, named after Barnsley FC striker Kayode Odejayi, was born. All of the Nicholsons are life-long Barnsley FC supporters and Odejayi had become a firm favourite with them after scoring the winning goal in the team's surprise victory over Chelsea in the FA Cup earlier that year. Unfortunately, though, the celebrations were cut short when Roger slipped up on the llama's placenta, injuring himself quite badly in the process. He had to be carefully helped back home and was laid up for several months.

That incident aside, llamas are generally fairly unproblematic. Allowing them to spend time outside means they keep their coats clean, as they shed their wool annually and don't often need shearing. "We have had to trim the llamas teeth on occasion, but it's better to get them out in the field for a few months every year," says Robert. "That way they assist in keeping the hedges trimmed back and the fibrous vegetation helps wear their teeth down."

Llamas don't thrive outside during harsh British winters, so at Cannon Hall Farm they are generally kept inside when it gets cold. In April 2020, when farmer Dale turned out the llamas for a run in the sand paddock, there was a charge as all of the llamas galloped to their new summer home. Skidding around the corner from their winter barn, the gang of llamas sped into the sand paddock and skipped around sizing up their summer accommodation.

Bringing up the rear was little Philippa the gllama, the pygmy goat who prefers to live with the llamas than other goats. The farmers tried putting her back with the goats, but she makes her feelings known that she doesn't want to be a normal nanny and always finds her way back to the llamas. She even had Priscilla the llama near her as a birthing partner when she gave birth to her little kid. For a time, there was a worry about what Philippa's baby would look like – had her obsession with hanging out with the llamas gone too far? Fortunately, when it was checked out, it was shown to be 100 per cent goat. She'd obviously had a moment with a Billy goat at some stage.

7

Fern and Ted at the Great Yorkshire Show

In the UK farming world, county shows don't come any bigger than the Great Yorkshire Show. It is one of the highlights of the farming year, a chance for farmers to enjoy a rare day off, mixing with other farmers and showing off their best animals and farming skills.

County shows happen all over the UK over the course of the summer, but the Great Yorkshire Show is considered by many to be the very best – the Oscars of farming.

The celebration of farming, food and country craftsmanship was founded in 1837 as an exhibition of farming stock and to promote agriculture. Originally held in York, it then moved to other locations in Yorkshire before settling in Harrogate in 1951. At that stage, around 54,000 visitors came to the show and each year it's got bigger and better, now attracting three times as many people of all ages.

"Richard, Robert and I have been going to the Great Yorkshire Show from the age of around five years old," says David, "so we all have so many fond memories from being there. I think we were even allowed a day off school to go to it because it was part of rural education and we all absolutely loved it. The three of us would always head to the fishing section first, but there was always loads to see and do, and hundreds of other kids, too. It was always a fantastic day out."

Back when they were children, the Nicholsons weren't exhibiting any of their own livestock and hadn't entered any farming competitions for many years. Roger had told them about his experiences competing with his animals at country shows when he was younger, but he had put showing on the back burner as he was more concerned with trying to make ends meet.

"There had been a 60 year gap for us in the show ring since Sam the beer drinking bull," says Robert. "So in 2019 it just felt right that we should get back into showing and try and recreate a bit of family history."

In February 2019, an episode of *Springtime on the Farm* followed Roger and vet Peter Wright on a trip to an auction in Oban in Scotland, where they bought Fern, a stunning four-year-old Highland heifer who was in calf. When the programme was aired later in spring, viewers saw Fern at the farm, looking as gorgeous as ever, along with her beautiful fluffy calf, who looked so much like a teddy bear he had been named Ted.

"From the very first time I saw Fern I thought she would be good enough to show in the Best Heifer Class," says David. "But Dad said, 'No way, look at the state of her: she's a tangled mess!'"

He had a point. While Fern was a magnificent example of her breed, she didn't exactly look like she had just stepped out of the salon. To maximise success in the show ring with the judges, she would need a bovine makeover. An animal must look its absolute best, and for a Highland cow that means a well clipped and groomed coat, trimmed hooves and polished horns.

Luckily, as David's wife Anita is a hairdresser, she had plenty of tips on how to get Fern looking catwalk-ready. "It took two weeks of around an hour a day's careful combing through Fern's coat just to detangle her," says David. "We didn't want to lose too much of her coat because we wanted her to look like a proper well-covered Highland cow. The trouble was we'd let her get a bit matted when she was out grazing in the fields. You don't normally groom cattle if they're out grazing. It's not a beauty contest out there, they just do their own thing."

Roger was convinced that the brothers would never get Fern ready in time, but that only made them even more determined to show him that they could. As well as looking the part, Fern also needed to be trained to act the part, walking calmly with her head held high. "Dad wrote us off when he saw me trying to walk her and Ted one day," says David. He said, 'You're a shambles!'"

With 10 weeks to get show-ready, David worked hard with farmer Ruth to train Fern to be able to walk alongside a handler, appearing poised and elegant, and then stand squarely for the part of the competition when she would be judged. It was also important to get Fern and Ted used to lots of noise around them so that they would not be distracted when they got into the show ring. Ted wasn't going to be judged on the day, but having him there could help Fern's chances, as the judge would be able to see what fine offspring she was capable of producing.

Come the big day, the family was on the road at four o'clock in the morning, so that they would have plenty of time to get Fern and Ted settled and ready for their category to be called. In the stalls where the cattle are prepared, the atmosphere was buzzing with hairdryers full on, last-minute knots de-tangled and various beauty products put to novel use to put the finishing touches in place. And with a touch of nail lacquer on Fern's hooves and a splash of baby oil to polish her horns, she was show-ready. Ted's lovely fluffy coat was brushed up and David and farmer Ruth got themselves smartened up ready for the competition to begin.

As if their nerves weren't already bad enough, the Cannon Hall Farm team's progress was being filmed for a television special all about the Great Yorkshire Show. The Nicholsons were used to the cameras, as they had already appeared in two series of *Springtime on the Farm* by this point, but competing in the show was a whole new ball game. Without the experience of competing in several smaller country shows, was it too ambitious to think that they could make the grade?

"The 'ring' is around 40 metres by 20 metres, and in the next paddock there's another class being shown, and in the next one another, and so on," says David. "I was really, really nervous and completely out of my comfort zone."

Finally, it was Cannon Hall Farm's time to shine and David and Ruth guided Fern and Ted out of the competitors' stall and into the show ring, while Robert and Roger watched anxiously from the sidelines. The proud expression on Roger's face said it all. Reliving his own experiences of exhibiting at the Great Yorkshire Show all those years ago, he could see the likeness of his father Charlie in his son David's eyes.

As David walked Fern around, a fellow competitor gave him a tip about keeping her steady. "I was holding her nose clip too tightly and he showed me how to do it, which was really nice of him – such a good sport. I was so nervous, though. I was concentrating so hard I could actually hear my own heartbeat."

Not that David knew this at the time, but the judge told the television cameras: "I'm looking for a workmanlike cow with a good frame, a good udder and good feet."

With her head held high, Fern looked very special and all those long hours of preparation certainly paid off. The judge looked more closely at each cow's physique from nose to tail, seeing how they carry their heads, the length of their body and their strong sturdy legs. It was an anxious few minutes, but then, all of a sudden, the judge pointed out David and he was awarded the first prize – a certificate for him and a red rosette for Fern. David turned to the television cameras and said: "Excellent. It doesn't get better than that, does it?"

David thought that after the judging he was free to now go and celebrate in the beer tent, but then he heard his win meant he was automatically entered into the Best of Breed category. "The trouble was I forgot my number, which could have been the end of the story. The Scottish judge said, 'It was a tirrible error whin he forgot his numberrrr, but it wasn't the cooo's fault, so I couldn't penalise him'."

In the end, Fern was awarded Reserve Champion (second place), which was an incredible end to the family's first competition attempt at the Great Yorkshire Show in 60 years. "I would have settled for anything, to be honest," laughs David. "When I think of when we started scrubbing up that tangled mess of a cow, who would have thought we'd do as well as we did!"

As for Roger, he certainly had to eat his words: "I was extremely concerned that we'd make a fool of ourselves, but Robert, David and Ruth proved me wrong, and I admit it. I'm so proud of them."

There were more reasons to celebrate in spring 2020, when Fern gave birth to another gorgeous calf called Bonnie.

Two days after the delivery, joy turned to concern as it became evident that the new calf wasn't feeding properly, and the brothers began to wonder if they would have to hand-feed her for the next few months. Their first priority was to check that Fern actually had some milk for Bonnie to drink and that everything was as it should be in the milking department. With a cow Fern's size, looking closely at any area underneath the body can be tricky, so they put her in a cattle crush in order to minimise the risk of anyone being injured during the examination.

With Fern safely installed, Robert and David tested her udders and established that everything was in working order. It was then just a matter of manoeuvring little Bonnie on to the teat, so that she could get the idea of feeding from mum. Things seemed to be going well, and so they removed Fern from the crate – at which point Bonnie promptly resumed trying to feed from the back of Fern, rather than from the side as she should. With Fern weighing in at over 600 kilos, this was highly dangerous. If Fern were to step on Bonnie, it could be fatal.

Happily, with more patient training, Bonnie finally got the hang of feeding from the side of Fern and the farmers could breathe a sigh of relief. Sometimes nature just needs to be given a helping hand.

8

More showtime fun!

Following Fern's magnificent success at the Great Yorkshire Show, the farmers decided to try and keep up the winning streak with a trip to Penistone Show.

The prestigious South Yorkshire event was first established in 1873 and, although not nearly as big as the Great Yorkshire Show, is said to be the largest one-day show in the north of England.

Competing in the 2019 Penistone Show it was now time for Roger to get back into the show ring himself, and everyone was rooting for him.

Fern needed a major clean-up when she arrived at the show ground – maybe she sensed the family's collective nerves? – but when the time came for her to be judged out in the paddock, she once again proved she was a superstar, and she and Ted took first place in the Champion Highland Cow and Calf class. And if that wasn't enough, she then went on to clinch the Supreme Champion prize of day and headed up the Parade of Champions. Roger couldn't have been happier. "It's over 60 years since I've competed in the show ring, but it feels like I've never been away," he said.

Following their 2019 success, the family was itching to get back to the Great Yorkshire Show in 2020, but when the COVID-19 pandemic hit, the show was cancelled and instead a virtual contest was arranged. "We'd been looking forward to the show all year," says Robert. "But an online show is better than no show at all – and luckily we'd carried on prepping Ted as if he was going to the Great Yorkshire Show. We'd kept his feed levels high and kept him tidy, but he needed a good shampoo to be ready for the big day."

Ted's winter layer was combed out, leaving a softer summer layer. Robert tried out a few hairstyles – including the Donald Trump look – then David gave him a shampoo. "My wife Anita has given me some nice herbal shampoo for Ted and I think he's enjoying it." Then David got ready for showtime again.

The judges needed to see photos and a video of Ted being paraded, so David and Ted headed down to the fields and Robert got ready to capture them on camera. Once uploaded on to the official judging site, they just had to wait for the results.

Unfortunately, it wasn't a repeat of their 2019 success. "Ted only came 5th, but he's a winner in my eyes and he'll have the chance to shine on another occasion," says David. "We all really enjoyed it, though, and we're really proud of him, whatever the result."

9

Emma, Hettie and Dougal

They may not be as famous as their Scottish herd mates, but Highland cows Emma, Hettie and Dougal are further examples of the fantastic breeds that live at the Nicholsons' farms.

"Emma won the same class at the Great Yorkshire Show in 2018 as Fern did," Robert explains. "The following summer, when we made our showing debut with Fern and Ted, we met Emma's owner, Dexter, who was the owner of the Supreme Champion cow when Fern got her Reserve Champion rosette in the same class. A while later he got in touch to tell us he was selling Emma and asked if we would be interested in buying her. It wasn't that he didn't love her – he was in the process of changing breeds on his farm and he wanted her to go to a good home."

And what better home could there be than Cannon Hall Farm, especially as the boys clearly knew what they were doing with Highland cattle. And there was an added bonus when they bought her: Emma had Hettie her calf afoot and was in calf again.

Strangely, Fern and Emma calved within 12 hours of each other – Fern giving birth to Bonnie and Emma giving birth to her gorgeous calf Dougal, who looked just like a teddy bear. "The fact that they are so synchronised is incredibly useful for us," says Robert. "With them coming into oestrus at the same time, by which I mean being ready to be mated, we have been able to get them served on the same day."

But no trips to other farms to meet handsome bulls for these two. They were impregnated by artificial insemination, bought in frozen batches from the Highland Cattle Society. "We buy 10 straws of semen frozen in liquid nitrogen and it's stored until we need to use it," David explains. "Then Philip, our contractor, defrosts it, puts it in a syringe and gets his arm right inside the cow and finds the exact spot he needs to release it. It's incredible, really, that you can freeze something like this and bring it back to life – and it's certainly more comfortable for the cow than having a huge bull on their back."

Hettie is strong and healthy and living her best life out at the Nicholsons' other farm, Mill Farm in the Gunthwaite Valley. "She's out there munching on the grass of the Penine foothills," says Robert. "It's not quite the Highlands of Scotland, but she's certainly prospered."

Meanwhile, Fern and Emma's new calves are due towards the end of April 2021 – around the same time as David's birthday – so there will be lots of reasons to celebrate!

10

The goat gang

With their compact cuteness and friendly bouncy personalities, pygmy goats are one of the visitors' favourite breeds at Cannon Hall Farm. Both children and adults just can't get enough of them. But long before the farm introduced stars like Millie and Primrose to the world, generations of goats of all shapes and sizes made their mark on Cannon Hall Farm.

"The first goat that we had on the farm was a white Saanen goat called Nanny," Robert explains. "I know it's not the most original name for a goat, but we loved her. She lived in the farmyard and we used her milk to rear lambs. But I wanted her for a pet. One day, when I was about nine, I decided to take her for a walk. I made a harness for her and coaxed her up the hill away from the house, but she was having none of it and high-tailed it back, pulling me over with her and dragging me along the tarmac. I kept hold of her 'lead' because it was all happening so fast and I tore massive holes in my jumper and trousers as she pulled me along. Mum wasn't best pleased and I learned a valuable lesson that day. Goats are not dogs."

The boys named their second pet goat Bubbles – after Michael Jackson's monkey. "He was a Saanen crossed with an Angora goat, so that meant he was very handsome and very tall!" says Robert. "He lived to be about 14, so he really was one of the family for many years. None of us tried taking him for a walk, though."

When Cannon Hall Farm opened to the public in 1989, the family borrowed some Angora goats, and shortly after they were joined by pygmies. The pygmy goats even got a mention in the local paper – Roger had thought he was buying just the one goat, but it turned out she was pregnant and she soon after gave birth to twins. From that point on, the goat population flourished and other breeds were introduced.

"We've 'professionalised' our goat breeding as we've gone along," says Robert. "Now we breed lots of very good quality goats that are sold on as pets and breeding stock. We also breed pygmies and Boer goats, including Stanley, who's a real farm favourite."

In the first series of *Springtime on the Farm*, viewers were introduced to Stanley as a tiny orphaned goat who was managing to hold his own amongst the newborn lambs. Fast-forward a year and the huge full-sized Stanley nuzzled up to David on camera. "David and Stanley have got a great relationship," Robert says. "However, they do bicker and row a bit and don't exactly see eye to eye." "I could never fall out with him, though," says David. "We had to bottle-feed him and he's one of those goats that we had to give a helping hand to, and now we wouldn't be without him."

Stanley is certainly one of the most affectionate animals on the farm and is particularly loved by young farmer Dale, who champions him and his inseparable pal, Gloria.

Gloria, an Anglo-Nubian, was initially bought as a dairy goat for milking demonstrations at Cannon Hall Farm, but she soon proved even more invaluable as a foster mum for orphan goat kids. She has also provided milk for many lambs whose mothers haven't been able to produce enough milk, and also pygmy goats who have to be hand reared. "She's been an absolute treasure to us on the farm, and she's done a brilliant job over the years," says Dale. "She doesn't always get as much attention from Cannon Hall fans as some of the other goats, but she and Stanley are such characters and they live very happily together. Stanley is also a great sheep herder – so he's very useful, too!"

Sadly, not all of the goat stories at Cannon Hall Farm have such happy endings. In the first episode of *Springtime on the Farm* in 2020, viewers saw Robert and David battle to keep newborn pygmy goat Gordon alive, carefully wiping it clean and clearing the tiny kid's airways by blowing into his mouth. Things looked to be progressing well as the bedraggled little billy took his first breath, but when the mum's maternal instinct didn't kick in, David and Robert had to tube-feed him and put him under a heat lamp near his mum, hoping the bonding process would naturally happen.

Gordon survived his first night and seemed to have picked up when Robert and David checked in on him again. But things still weren't right with the little mite and, following a check-up with the vet, he was discovered to have birth abnormalities. The farmers continued to hope that Gordon could survive, but 12 days later Rob's fears were realised. "I went to feed Gordon and instead of him coming running excitedly for his milk, he was in the corner, just looking like he was sound asleep. He'd died in his sleep, which was very sad."

"It's so tough when you lose an animal that you have so much invested in," says Robert. "With an animal like Gordon, he always had his challenges, but we believed we could rear him and that it was going the right way, so to lose him the way we did was a really big blow. I really thought it was a battle that we were winning."

Because of their size, when a pygmy nanny goes into labour, it's not as easy for farmers to help with their delivery. If a birth is complicated and the farmer needs to adjust the baby kid inside the mum, there's very little room for manoeuvre. You have to hope that any new kids are positioned well for delivery – especially when it comes to twins.

When pygmy twins Charles and Camilla were born, Robert and David also feared for their future. "We knew that there wouldn't be enough milk for both of them, as her mum only milked on one side," says Robert. "The little girl was spending all her time sucking a teat that didn't work and burning all her energy for no reward. We knew her only chance was for us to become mum to her."

Viewers, farm visitors and social media fans instantly fell in love with Camilla, or Millie as she came to be known. Robert and David moved her to their special care baby section at the farm, where she was bottle-fed every two hours. "It was unlikely that she got any colostrum from her mum, which is what's really needed in the first few hours of life to support her immune system, so we had to watch her very carefully."

Millie even had her own soft toys to keep her company. "It wasn't a foregone conclusion that she would survive. Like Gordon, she had some big challenges to overcome," says Robert.

With regular weigh-ins showing tiny increments of weight gain, things started to look more positive for Millie. And one day, there was even more good news. A viewer called Amanda got in touch to say that she had an orphaned pygmy goat called Primrose that could potentially be a playmate for Millie. Robert was sold on the idea from the moment he clapped eyes on beautiful little Primrose; he could tell straight away that she would be the perfect pal for Millie.

After a couple of days of quarantine for Primrose back at Cannon Hall Farm, Robert and David built her and Millie a mini-playground complete with hay bales and a slide. Within minutes of being introduced to each other, the two were sliding around and were the best of friends. "Companionship is so important for pygmy goats," says David, "and as the two little orphans potentially could have compromised immune systems, putting them together meant they could be weaned together without fear of picking up any infections from any of the other goats."

The majority of the goats at Cannon Hall Farm live indoors, as they seem to prosper better and are less prone to worms that sheep carry in the fields. They also love their specially built adventure playground that Robert and David created for them in 2019. "It's important that goats have as much enrichment in their lives as possible," says Robert. "And as we can't graze them as much as we'd like to, they have a playground to climb and jump around on with all their friends to keep them happy."

11

The Magnificent Shires

One of the earliest photographs of Roger Nicholson shows him sitting proudly on top of Blossom the shire horse, so it is fitting that shires still have a very special place at Cannon Hall Farm.

Long before the days of tractors and combine harvesters, shire horses did the lion's share of farm work, pulling ploughs and carts, working the land and more than earning their keep. Their sheer size meant they weren't a cheap animal to keep – historically, shires have held records for being the world's heaviest and tallest horse. They are incredibly strong and effective working animals, but like all horses, they need special care to ensure that they are kept in good condition.

After the Second World War, huge changes in agriculture meant that cumbersome expensive shires were replaced by machinery, and their numbers began to decline rapidly. By the year 2020, there were only one and a half thousand breeding mares in the UK.

The first mighty shire at Cannon Hall Farm – coincidentally also called Blossom – was borrowed from the Nicholson brothers' cousin, Richard. She was such a hit that the Nicholsons decided the time was right to have a pair of shire mares of their own.

"The original shires that we owned were called Poppy and Lottie," says Robert. "Around four years ago, Poppy gave birth to a foal, but unfortunately the birth was very difficult and Poppy never got back up again and had to be put down. Then, Chester, her foal that we were hand-rearing, got sepsis and he died suddenly, so initially we had a tragic history with shires."

It took three long hard years for them to get over the loss of their beloved Poppy. "Whenever you have heavy horses you always have jeopardy," says Robert. "There's always a risk with everything you do because of the size of them. If something goes wrong, like a fall or a slip, it tends to be a big fall or slip – and that's the danger. Any kind of mishap puts a massive strain on their muscles and ligaments."

Taking the plunge with a new mare who was both pregnant and already had a young foal with her, the Nicholsons took delivery of a new shire called Orchid in early spring 2019. "We bought her from a farm in Shropshire and the fact that she had a filly afoot was a real bonus," says Robert. "We were getting two fillies in one go, plus she would probably be giving birth during the time that *Springtime on the Farm* was being filmed. So she was very much a considered investment."

The young filly that came along with Orchid had been sired using artificial insemination from a champion shire stallion. She was every bit as gorgeous as Roger's childhood shire horse Blossom, so she too was called Blossom. Shires traditionally have very good temperaments and Blossom proved to be just as friendly, particularly with resident equine expert Ruth.

When Blossom could be weaned from her mother, Orchid, farmer Ruth took time to closely bond with the young foal. "All the stroking, feeding and talking that I did with her helped me to train her. We teach all of our horses good manners and so by bonding with her it encourages a natural trust."

Blossom may have got thumbs-up all around for her gentle temperament, but her mum Orchid quickly got the nickname "Awkward Orchid" when she arrived at Cannon Hall Farm. "To be fair to her, when we first got her, we were told that she didn't like having her feet done," says Robert. "We were warned that she'd push and shove."

"If Orchid doesn't want to do anything, she won't do it," adds Ruth. "She's very headstrong, but then she's so gentle as well."

Awkward Orchid certainly lived up to her reputation. On her first encounter with Dan the blacksmith she refused to lift up her feet and needed to be sedated. "We have to get the vet out each time she gets her feet trimmed so it ends up being a very expensive job," says David. "But it's not fair on the blacksmith not to have her calmed down, though, because she literally weighs a tonne and if she decides to stamp on him it would be very dangerous."

"It's amazing how powerful they are," says Dan the blacksmith. "She's not nasty, but she could throw me around like a rag doll if she really wanted to."

"Orchid is actually a really stoic, well behaved animal," says Rob. "She loves her food. She's a strong independent woman and we think the world of her. We do need to trim her feet, though, because over winter, when the horses have softer bedding, their hooves grow longer, so they need trimming five or six times a year to stop their feet getting waterlogged or infected."

When the time came for Orchid to have her foal, Robert and David's mum Cynthia was keeping an eye on the barn CCTV, which alerts the family to anything happening with the animals when the farmers aren't in situ. Cynthia noticed that Orchid's waters had broken and phoned Robert and David, who then sprang into action and ran to be near her. "We didn't want to interfere with the birth, so we just waited around the corner from the barn and watched the CCTV on my phone." Robert explains.

Horses often give birth at night, a natural instinct as there's less chance for predators to harm their foals. Births generally take between 15 and 45 minutes. "She laid there for a good 10 to 15 minutes with the foal's front feet hanging out of her," David remembers, "then she got up and delivered him quite easily. We wouldn't have intervened with the birth, as that can spook the mother, but we could see that the membrane sack around the foal hadn't broken, so I quietly nipped in and broke it around its mouth so that it could breathe easily. It was exciting and worrying at the same time – especially after what had happened with our first mare, Poppy."

"Luckily, Ruth had briefed us both on what the birth of a shire horse should be like," says Robert. "When the foal is born, blood is transferred from the mare to the foal via the thick umbilical cord, so the foal has all the blood it needs in its body. We knew we had to leave them both there nice and quietly while this process happened. Orchid lay there motionless for about two minutes and, just as we were starting to get a little bit worried, she finally stood up."

"We then noticed that the foal's back leg was wobbling a bit. He'd been cooped up in his mum's womb for so long that he was a bit like a drunken man and couldn't get his coordination right – he was like an antique table," says Robert. "So we milked Orchid and gave it in a bottle to the foal, as he wasn't stable enough to latch on to her straight away."

After a check-over from the vet, everyone was reassured that everything was fine and the team could breathe a sigh of relief. The following morning, the foal was looking as fit as a fiddle. "We decided to call him Will, in tribute to Will Roe, who helped our father 60 years ago just after he had lost his dad. He was a local farmer and he stepped in and made a world of difference to Dad."

But even before visitors and television viewers were introduced to Will, there was another equine arrival in July 2019, when the eight-year-old shire horse Ruby arrived at the farm. "Robert said to me, 'I don't think we have enough shire horses, we should buy another!' And off he went!" David recalls.

"*Springtime*'s presenter Adam Henson told me that there are only 300 shire foals born a year, so we felt we should do our bit to help the breed," Robert explains. "They're part of the past, but they can be part of our future because they are so iconic. Places like Cannon Hall Farm should be at the forefront of promoting their attributes."

As soon as Ruby and Lottie met in the beautiful sunshine, they were immediately firm friends. It was a heart-warming sight to watch them trotting around the paddock together and Lottie was obviously a little bit in awe of the new arrival.

Having had time to bond with each other and settle in, it was time to start thinking about breeding. "Mares are in foal for the best part of 12 months, so it's a long-term undertaking and a big commitment," says Robert. "It's also very expensive. If you want a really high-quality foal, it's definitely not something you can do on the cheap."

Wanting to get the very best match for their shires, the Nicholsons selected a premier stud farm near York, home to national shire champions. "Ruby is a very traditional type of shire horse with a great pedigree, so we wanted her to be matched with a very special horse," Robert explains. After meeting the potential suitors, Macbeth (or Big Mac as he is known at the stable) was the stallion chosen to do the honours. Not only was he a magnificent looking horse, but his owners had tipped him as a future national champion.

And it couldn't have had a better result. At the end of January 2020, it was confirmed that Ruby was pregnant. Would it be a colt (a boy) or a filly (a girl)? "Ideally, we want to always have filly foals and then she can carry on breeding," Robert explained. "Ruby's bloodlines are terrific and a filly foal would mean that we could build our shire horse future around her."

To the brothers' delight, in August, she gave birth to a beautiful filly foal that the boys decided should be called Sapphire. She was a little wobbly on her feet but soon got the hang of feeding, and a couple of weeks later was proving to be mischievous, cheeky and feisty. Bouncing around her paddock with boundless energy, she also had a habit of kicking out with her back legs – as farmer Ruth became only too aware of. "She was very headstrong for a filly foal, which is quite normal," said Ruth. "Her mum Ruby is placid and quite laid back and we were hoping that she would discipline her foal a little bit, but it just didn't happen."

Nevertheless, with Ruth's skill and determination, like her name suggests, Sapphire turned out to be a true gem.

During summer 2020, Orchid and Will went off on a little summer holiday to the same stud farm where Ruby and Macbeth had got together. As Will was still such a young foal, he needed to still be close to feed from mum.

David chose a stallion called Double Trouble to be the horse that would hopefully be the sire to Orchid's foal. "Even though he wasn't the national champion, I loved the colour of him and I thought it would be lovely to have a pretty grey foal."

Leaving them at the stables while Double Trouble and Orchid got together meant it was an emotional goodbye for Robert, who had grown very fond of Will the foal. "I'll miss Will a lot," he said in the episode of *This Week on the Farm* that saw the boys taking Orchid away for her holiday romance. "Every day I go and give him a back scratch and it won't be the same without him. I know that one day he'll have to go and make his life elsewhere, but at the moment we want him back healthy and happy and we want Orchid to have a foal next year."

But for some reason it just didn't happen. Double Trouble had plenty of attempts to get Orchid pregnant but the chemistry wasn't there. "Sadly, Double Trouble didn't get Orchid into trouble," says David. "They went on lots of dates, but it just didn't happen – and Rob will never let me forget that I got it wrong!"

A different plan of action was needed, and Robert and David selected Charlie, the national shire horse champion, to mate with Orchid, hoping that the investment would pay off. Three months later, the brothers went back to the stud farm in York to be reunited with Orchid and Will, who had by now grown into a very fine teenager. Robert was overjoyed to see Will and greeted him saying, "Hello, remember me, we used to be best pals and we can be again. Want a back scratch?"

In order to find out if the meeting of Orchid and Charlie had paid off, Robert and David tested the water with their feisty Shetland ponies Jon Bon Pony and Ozzie Horseborn. Don't panic, the little stallions weren't there to try and mate with the supermodel mare. "If she is in season, she would lift her tail as a way of showing that she's attracted to him. Not that we would ever mate them, of course, it's just a basic way of working out if she is pregnant. "If they look like a stallion, smell like one and sound like one, even if they are half the size, they will have the same effect," Robert explains.

Results were inconclusive as both Shetlands and shires looked a little confused about their blind date. It was only when Shona the vet did a proper ultrasound test that she was able to confirm if Orchid was pregnant. Sadly, the news wasn't good: she wasn't carrying a foal.

It was a bitter blow to Robert, David and Ruth, but they were philosophical about the situation. "We're not too disheartened because it is sometimes difficult to get a suckling mare back in foal straight away," Robert says. "So we're going to give her a rest and take her back to the stud early next year and have another go. We really think she has too much quality not to try and get another foal from her."

And as Orchid looked forward to a restful autumn, her young foal Will was on the move again. Having been so impressed with the young stallion during Orchid and Will's summer holiday, Bill from the stud farm in York wanted to buy him. "It's with a very heavy heart that we've sold Will, because I've really enjoyed his company and I'm going to miss him," says Robert. "But it's a great joy that he's going to such a good home. He's been bought as a potential stud, but if he doesn't quite cut the mustard as a stud, he will make a top-class riding horse, so he's got a great future ahead of him."

Throughout the summer shenanigans with Orchid and Will, Lottie, the original shire at Cannon Hall Farm, was going through pregnancy dramas of her own. Sadly, she lost the foal that she was carrying in 2019, so she also went to visit the same stud farm in York as her fellow mares. This time everything went to plan and a new shire should be coming to Cannon Hall Farm in spring 2021.

Then again, of course, there's always a chance that in the meantime Robert will buy another one…

1 2

Spike the Porcupine

Rodents are one type of animal that don't usually get a red-carpet welcome at a farm, but then again, as we all know, Cannon Hall Farm isn't exactly a traditional farm. When it first opened to the public in 1989, mice and rats were celebrated in the Mousetown enclosure, but there are much more spectacular rodents now living happily at the Barnsley farm.

Spike the porcupine has been living at the farm since 2019. "When we got our zoo licence and we were legally allowed to keep more exotic creatures, we thought, 'Why not have a porcupine?'" says David. "As kids, we always loved fishing, and porcupine quills are often used for floats, so we've always had a fascination for them. For a long time, it was at the back of our minds that if we could ever find a porcupine, we'd try and make a lovely home for it here on the farm."

The Cape Porcupine is the world's largest variety of porcupine and is usually found in central and southern Africa. Spike, however, came from Birmingham Wildlife Conservation Park.

"He's a magnificent creature, but he was very, very shy when we first got him," says David. "It was difficult to coax him out of his shelter, so we started to put his food at the bottom of the ramp in his enclosure so that he would venture out to fetch it, and gradually he got more and more used to seeing people around him."

Farmer Darrell discovered that there's nothing that Spike likes more than a potato to snack on, and he has been able to capture Spike on video for the Cannon Hall Farm website by luring him with his favourite treat. At first, Spike would do little more than poke his head out of the shelter, but Darrell's patient training has paid off and while he is cleaning up his cage, Spike is happy to come out and demonstrate some posturing and quill shaking.

With their thousands of long hollow spiny quills, porcupines in the wild rattle their quills to scare away predators. If cornered, they will viciously charge at an attacker, flexing those quills into sharp weapons. Thankfully, Spike is much more relaxed now that he has worked out that he isn't in danger. "Porcupines are very good diggers and when Spike first arrived he had a good crack at trying to dig his way out, but he has never shown any aggression towards anyone," says David. "When I'm working on the farm late at night, I'll often visit his enclosure and have a little chat with him. I've tried to make several videos with him for the website, but he just runs away!"

Being so camera-shy is possibly the reason why Spike hasn't yet become one of the Cannon Hall TV stars, but all that may change. In November 2020, Spike was joined by a female porcupine called Winnie and the team have high hopes that spiky new babies might come along one day.

Cape porcupines mate throughout the year with gestation periods of around three months. They have litters of up to three babies, which, thankfully for their mums, are born with soft quills. The great news is that porcupines can live for up to 20 years in captivity, so hopefully Spike and Winnie will be at Cannon Hall Farm for a long time to come. Let's just hope the two giant rodents rub along nicely together – and quite literally don't get each other's backs up.

1 3

Jon Bon & Ozzie Horseborn

He's perhaps the biggest animal star of the show, even though, at just three feet tall, he's not even half the size of most of the other equines at Cannon Hall Farm. But what Jon Bon Pony lacks in height he more than makes up for in personality.

There are no prizes for guessing that Shetland ponies originated in the remote Scottish Shetland Isles, where small horses have been kept since the Bronze Age. The harsh climate and scarcity of food in the northern islands meant the ponies developed into extremely hardy animals. With compact bodies and a long mane, it is easy to see why everyone loves the breed – and Jon Bon Pony in particular...

"We bought Jon Bon in May 2019 from a farm in Aberdeen," says Robert. "He was 11 years old and he looked to be in the prime of his life. He was pumped and ready for action, and David and I knew he was 'the one'."

Prior to the arrival of Jon Bon, the brothers had borrowed a Shetland stallion to breed with the Shetland mares that lived on the farm, but with little success. Artificial insemination was an alternative option, but the brothers decided to take the plunge and buy a little stallion of their own.

"We could have bought a stud that was younger than Jon Bon," says Robert. "But we reckoned he looked the part – he was prancing around and making all of the right noises – so we were sold. He was great value, too – just £400 for a pedigree animal. We also bought two more new mares for him to breed with."

Back at the farm, the brothers took to social media to find a name for him. As usual, online friends didn't disappoint and came up with plenty of great suggestions, but Jon Bon Pony was the clear winner. "With his thick mane of hair and that rock-star swagger, it was perfect," says David.

And as soon as Jon Bon Pony was released into the pen with his adoring mare groupies, it was clear he had an eye for the ladies and started whinnying to them. With their big bushy tales, mares often need a bit of help in the mating department. "Sometimes it's a matter of holding their tails to one side, or we bandage the tail up so that access isn't denied," says David. "I was trying to help him along, but he kicked me twice, so I thought, 'Okay, mate, I'll leave you to it'."

Feeling confident that Jon Bon had earned his keep and serviced all of the mares, the brothers called in the vet to find out if there would be a pattering of tiny hooves in spring 2020. Using ultrasound, the vet scanned all of the mares, but found that only one of them was pregnant. "It was too late in the season to try again with the other mares," says Robert. "We knew it was a case of waiting for the following spring to see if we could rekindle Jon Bon's ardour, as spring is when horses naturally ovulate."

The brothers were concerned that there might be an issue with Jon Bon, as they knew the mares were in good health, and they began to get seriously worried a short while later when they noticed that his superstar shine seemed to have dimmed a little and his handsome head that he normally held high dropped despondently. "He really looked like a horse in decline," says Robert. "He was losing weight and losing condition and it's been sad to watch. He's normally such a feisty character but he'd become a real pipe and slippers man."

The vet checked Jon Bon over for possible reasons for his weight loss, thinking it could be down to a stomach ulcer, or possibly something worse. But after giving Jon Bon a thorough examination, a close dental check showed he had a problem at the back of his jaw. His upper and lower sets of teeth weren't fitting together correctly, and he had developed a nasty hook on his back tooth which was causing him a lot of pain. No wonder he had been off his food.

Using a special halter to ensure his mouth stayed open, Mick the vet filed down the problem tooth with a giant version of a dentist's drill. "It looked quite basic, but I think that without it we would have lost Jon Bon because he was losing condition at a rate of knots. He looked like a shadow of the animal that we had bought just a few months earlier. He just seemed to have lost his lust for life – he was like Eeyore from Winnie the Pooh."

A week later, Jon Bon had his sparkle back and was eating like a horse again. As David the vet said: "He's no longer living on a prayer." (Warning: there will be more awful Jon Bon Jovi jokes to come...)

While the brothers were chewing over the issue of Jon Bon's tooth, the little stallion's girlfriend, Pony Em, was having troubles of her own with her baby foal. "We saw from the CCTV footage on the farm that he was born about midnight, but he wasn't feeding properly," says David. "It was vital that we got his tummy full of his mother's milk to get antibodies to fight infection and energy."

Finally, after lots of perseverance from farmer Ruth, the little foal that they named Pony Hadley started drinking naturally from his mum. Soon the pretty little black cutie was well on his way to being every bit as popular as his dad. And when the news reached the former Spandau Ballet singer Tony Hadley that he had inspired the little foal's name, he sent a video message to *This Week on the Farm*: "I'm absolutely honoured that the foal was named after me, which is brilliant news – solid Gold! If ever I'm in Barnsley, I'll come up and see the little guy, and I hope he's doing well!"

Once Jon Bon's teeth had been sorted out, Robert and David hoped that his natural charisma and love for the ladies would soon mean that he was back to his old self again, but sadly that was not to be. "At one time, if I walked past his pen with one of the shire horses he'd be whinnying at the pen and knocking at the gates saying, 'Let me at her, she's not too big for me, bring it on!'" says Robert. "But Jon Bon was giving love a bad name," adds David. (You were warned about the Jon Bon Jovi jokes…)

After one last chance with a Shetland pony mare that was in season (and a bit of comical coaxing from Robert – "Come on, she's here, isn't she pretty?"), Robert and David decided to put plan B into action. The breeder who sold them Jon Bon suggested that a spell of isolation might make the heart grow fonder, so the lacklustre pony was taken to a private pen and given a luxury diet to lift his spirits. "We gave him oats, barley, wheat, molasses and carrots – the food of champions," says David. "We really believed that recipe would do the job. We wanted him to rest, to eat the best food, to come out and be the Mr Lover, Lover, Mr Boombastic that we know he can be. We were so hopeful that after three weeks on his own it would happen."

"We felt bad taking Jon Bon away from his ladies," says Robert. "But we took expert advice from the breeder we had got him from. The breeder had been around Shetlands all his life, so we knew we were doing the best we could for him."

Before the previously lethargic lothario was let back with the ladies, Robert and David treated him to a luxury makeover. "He looks like an old carpet and we want him to look like a Persian rug!" Robert said as he groomed his little friend. Finally, it was the moment of truth and Jon Bon was paired up with the feisty pony Priscilla. It was a positive start, but Jon Bon turned his back on her advances.

Sadly, it was to be the end of the line for Jon Bon's days as the Shetland farm stud. "He was neighing and stamping and getting excited, but then, nothing. I think he was just happy to see his friends again, but that was it," Robert says. "You can lead a horse to water..."

"Maybe if we had put some sexy Barry White music on it could have worked," says David. "But really he wasn't even halfway there. But we were determined that whatever happened, he'd always have a future with us – even if it was a non-breeding future. Who knows, maybe our little rock star will have a comeback tour one day?"

Of course, that wasn't the end of the story, as the boys were still very keen to have Shetland pony foals being born the following spring. So off they went to Scotland to visit a Shetland pony breeder in Ayrshire. "12-year-old Unicorn had been abused in a former home, but Elsbeth, the breeder, had rehabilitated him and trained him to trust in people again," says Robert. "I'd seen a picture of him and he looked just as good in real life. He was a bit nervy, but Elsbeth was extremely confident that he was a breeder. In fact, she said, 'If he doesn't mate with them I'll eat my hat!'"

A new life calls for a new name. "He looked a proper little rock star, but we didn't feel that he had a rock star name," says David. "I thought maybe Pony Christie, but that sounds more like a crooner's name, then Robert came up with Ozzie Horseborn, which we both agreed was a winner."

Ozzie's new stablemates were also delighted with the new stallion on the block. "David held on to Ozzie in one corner and I had Alice, the most dominant mare in the other corner," remembers Robert. "We counted to three and then it was mayhem as Ozzie started mating the other mare that was in season. Alice then weighed in and started kicking him and we had to protect Ozzie while he was in action. We held her back while Ozzie concentrated on the task in hand and luckily Alice seemed happy enough getting second dibs."

"The pursuit of true love definitely came with danger, as he got a few knocks along the way. But very quickly we were pretty confident that Ozzie had got all of the Shetland pony mares pregnant."

At the time of writing this book, the little ladies hadn't yet been scanned – and as Orchid has proved, not every equine tale has a happy ending. But whatever happens, the Shetlands will continue to be farm favourites and there will always be a place in everyone's hearts for Jon Bon Pony.

14

The Reptile House

Along with the fluffy, furry and four-legged favourites at Cannon Hall Farm, there are also more than 25 breeds of weird and wonderful wild animals waiting to be discovered at the Reptile House.

As soon as you step across the threshold of this heated oasis in the middle of Yorkshire, you will be transported to the tropics.

For example, look up towards the ceiling and you will see a legion of half a million leaf cutter ants, busily transporting tiny pieces of leaf along a rope system suspended from the ceiling back to their nest. There they break down the leaves which eventually decompose into a fungus which they then feed on.

"Our entire leaf cutter ant colony is from just one queen," says Cannon Hall Farm's resident reptile carer, farmer Kate. "They're very self-sufficient and don't have to have any special leaves to eat, though I've discovered they don't like Holly. I think it's too thick for them to cut. But that works for me because it means I don't have to collect it to feed them."

As every Cannon Hall Farm fan knows, many farm favourites have their own names, but the team drew the line at naming each of the ants. Not so the iguanas, though, who were given their names by Kate. "The girls are Cleo, Martha, Penelope and Izzy, and they live with our adult Green Iguana male Iggy. He's a very happy chap with his four females."

Iggy was one of the first reptiles that came to the farm from a reptile rescue centre in 2017, along with Rex the female Asian water monitor. "One of my favourites from the rescue centre is Picasso the panther chameleon," says Kate. "He came as a juvenile and he had quite a bit of an attitude on him to start with. When I opened his cage to feed him, he'd hiss and open his mouth threateningly, but now he's very chilled out with me."

Having the ability to sense the daily moods of the animals in the reptile house is certainly very useful. "I wouldn't handle our big boa constrictors Frank and Fergus if I thought they were in a grumpy mood," says Kate, "but when you work with an animal every day you get to know their behaviour and what kind of mood they're in."

Luckily, the poison dart frogs, which are considered one of the world's deadliest creatures, carry no threat at the farm as they don't have access to the toxins they would eat in the wild. Nevertheless, all of the reptiles and amphibians are fed a rich diet of fruit, vegetables and, if their species requires, fish, mice and live insects.

"We do a yearly diet review, plus a weekly weigh-in to make sure that all of the animals are on track to see whether we need to change our feed plans," Kate explains. "For example, we have a bearded dragon called Sherbert (he has a yellow beard like a lemon sherbet) who is on a little diet at the moment."

Getting some creatures to behave on the scales isn't easy. "Iggy the male green iguana is so chilled out I can just pop him onto the platform and he will take it in his stride, but I have to bribe Rex and tempt her on the scales with some tasty fish or worms."

Two of the earliest residents of the reptile house were David's tortoises Itsy and Bitsy, and they still live happily in the Reptile House, often hand delivered cabbage leaves by David's wife Anita.

The main middle tank of the reptile houses is home to a curious looking soft-shell turtle called Piggle, who has a funnel-shaped nose like a pig. It's certainly not what you'd expect to see on a farm visit, but, as we all know, this is no ordinary farm.

Farmer Ruth has come to be known as the equine expert at the farm, but she is also a real expert in the "red tape" side of things and played a key role in ensuring that everything was in order at the Reptile House. When the farmers were applying for the zoo licence, for instance, one of the requirements was there was an escape plan for the Reptile House. Ruth explains: "This means that if an animal escapes, you have plans for how to keep the public safe, the staff safe and the animal itself safe. Obviously, other zoos work with elephants and tigers, but everything has to be standardised and we have to adhere to the same high standard for every single animal."

"Our big white Burmese Python, Apollo, is classed as our most dangerous animal and so we have lots of safety protocols if he ever got out," says Kate. "He is a constrictor and he has quite sharp teeth and, although not venomous, he could inflict a nasty bite."

Luckily, with health and safety at the forefront at Cannon Hall Farm, Kate has never had to find out how sharp those teeth are. "The worst bite I ever got was from a guinea pig!" she says. "When we are working with Apollo a few members of staff have to be present. We are always extra careful."

The Reptile House wasn't featured in the first few series of the television programmes filmed at Cannon Hall Farm, and as a reptile fan, Kate is keen to spread the word about how great they are. "Often adults are much more afraid of certain reptiles than their children are. But once they know more about them, they are often less fearful. Plus we have such great characters here."

As well as reptiles, there are amphibians, tropical fish and invertebrates such as hissing cockroaches and land snails. Each species has different dietary needs and is well catered for in a specialised area at the Reptile House. "Our zoo licence requires us to have separate meat and vegetable areas, and we use different knives, chopping boards and fridges that comply with the same rules and regulations for human catering," Ruth explains.

From chopping up veggies, defrosting frozen mice, cleaning out and checking the humidity of the tanks, to weighing and health checking every species, there is an incredible amount of work that goes on behind the scenes in the Reptile House to ensure every creature is living their best lives, safely and happily.

15

Chicken Run

When Richard, Robert and David were growing up, chickens were the first farm animal that they encountered. While there were other farm animals out in the fields, the boys loved chasing the family Bantam hens around the yard and helping their mum collect the eggs for breakfast every day. As they grew older, we heard earlier how they discovered money could be made from their pet chickens, charging 2p a stroke for the visitors to nearby Cannon Hall park, and selling random boxes of forgotten eggs.

"We've always had lots of poultry on the farm," says David. "As lads we used to hatch and incubate our own chickens and every year we'd breed turkeys at Christmas time."

In spring 1989, when Cannon Hall Farm first opened to the public, chickens were hatching in the incubators and there was a range of rare breeds and other poultry on show, including fancy pheasants, quail and crested pompom-headed ducks.

All that came to an end, however, in 1997, when bird flu hit the UK and the brothers took the difficult decision to rehome all of the farm's feathered friends. "We loved having lots of different birds, but back then we were much more of a hands-on petting zoo and we knew that an outbreak at Cannon Hall Farm would be disastrous," Robert explains.

Years later, when it was considered safe to do so, eggs ready for hatching were welcomed back for springtime. Roger's Bantams were rehomed in the family chicken shed, and in 2020 they were joined by some rather magnificent-looking Shetland chickens.

Robert and David became particularly taken by the breed after visiting a farm near York and seeing its collection of black and silver, cream and brown Shetland hens, which lay distinctive greenish-blue eggs. "We were quite used to handling cows, sheep and pigs," says Robert, "but we were a bit out of practise when it came to chickens, so I let Dave do the running around when we were filming them for an episode of *Springtime on the Farm*."

"I actually caught one! But they didn't show it on the programme," says David, obviously still smarting from the experience. "But instead of actual chickens we came back to the farm with a tray of 10 eggs to hatch, hoping we'd end up with some hard-working hens."

In March 2020, *This Week on the Farm* caught up with the feathered new arrivals. "It's been an absolute pleasure watching these chicks on a daily basis going from tiny little balls of fluff to what they are now," Robert said. "Just the smell of the poultry shed takes us back to when we were little."

As tiny chicks, they were happy under a heat lamp in the chicken shed near Roger and Cynthia's house, but once they had tripled in size, they were very definitely ready to leave the nest. Luckily, Robert and David had a rather nice new home lined up for them.

The swanky new hen house had lots of room to house the chickens at night, complete with plenty of ventilation to keep their home cool, plus nest boxes for laying their eggs. As free range birds, during the day they would have a spacious new realm to discover. "The benefits of letting them run free range are vast. They can explore, eat the vegetation and the insects, it's just a fabulous environment for them," said Robert. "It's been a long time since we've had chickens on the farm and it's great to have them back."

Even though all of the chickens were healthy and hearty, there was only one hen amongst them. "And it didn't end well," says Robert. "Unfortunately, the day after we moved all of the chickens into their new house, they were all outside having a peck around and a fox killed our one and only hen."

Nevertheless, the boys are happy that the cockerels are enjoying life at Cannon Hall Farm, and they are proving to be a big hit with visitors. "People like seeing them around and they enhance the look of the farmyard," says Robert. "There's a red one called Rupert that's my favourite and I like hearing the cockerels crow, so even though we haven't ended up with any hens, it's not all bad. It just means we'll have to try again – maybe with a different breed next time."

Percy Pig, Rocky, Lucky and their ladies

Sir Winston Churchill is rumoured to have once said, "Dogs look up to us. Cats look down on us. Pigs treat us as equals." Known as one of the most intelligent farm animals, pigs certainly have a special place in Roger Nicholson's heart and he is known as Cannon Hall Farm's resident pig expert.

"I wouldn't say pigs are my absolute favourite animal," says Roger, "but the pigs need me the most." Earlier in 2020, Roger's right hand pig man Nigel passed away, having worked at Cannon Hall Farm for 12 years. "He had such a rapport with the pigs. As the boars heard his voice, they would run to the door to meet him."

Roger is being modest, as it is obvious that he also has an affinity with his 45 sows, three boars and their numerous piglets (around 300 at any one time), not to mention Mildred, the farm's huge hairy Mangalitsa pig.

Having been brought up on a farm where his father nearly always kept pigs, Roger has learned a thing or two about porkers. Gone are the days when they would be fed pig swill – in Roger's childhood days, the swill was mostly leftover school dinners from the local school, plus barley and wheat from the farm, now it's a specially formulated diet of high quality milk and cereal pellets to keep them in peak condition. "It's not a good idea for the sows to get too big," says Roger. "Once they have given birth it's very easy for them to flop down and squash their piglets. So when they have little ones we want the sows to be fit and agile."

All the pigs live indoors at Cannon Hall Farm, as the land there is heavily clay-based. Outdoor pigs ideally need sandy soil to thrive. When it comes to farrowing (birthing) time, Roger prepares a soft bed for his sows with plenty of straw and wood shavings to keep everything comfortable, dry and warm.

Following a pregnancy lasting three months, three weeks and three days, sows give birth to litters of between 10 and 16 piglets, although Roger has on occasion seen a sow produce a single litter of as many as 24 little piglets. "Compared to lambing and cattle birthing, farrowing is like shelling peas and they're tough little competitors," he says. "The average birth takes about four hours, though sometimes it can last a day if they are in a bit of trouble, in which case we step in and give them a helping hand. But for the most part we just keep an eye on them while they birth their young."

He's clearly a big softie about his beloved sows. They even have underfloor heating. "When the piglets are being born it's quite a wet job, so they need to be bedded down properly so there are no bare areas on the concrete floor. And they love a good tummy scratch, which is also a great way to get a sow to lie down when she's just about to have her piglets."

Pigs need little help when they are farrowing and quickly find their way to the teat once they are born. Even the tiniest of the litter may not end up being the runt. "The runt appears when they don't get as much milk as the others," Roger explains. "When they have selected a teat, they stick to it and if it's not a very good teat and doesn't produce enough milk, then that piglet will be the runt of the litter. It's not necessarily the smallest one born. The smallest one may catch up and overtake the others if it's onto a good teat."

Yorkshire Vet Peter Wright adds: "Vets rarely have to intervene with farrowing. When we do, it's generally because one of the litter has got stuck in the birthing passage and a vet has to help them out. Like buses, you have none, then suddenly you get three at a time…"

Rather than referring to them by name, the sows are all numbered to avoid complication – there are 45 of them, after all – but the Cannon Hall Farm boars have been christened Rocky, Percy and Lucky.

You may remember Lucky the boar when he was featured in an episode of *Springtime on the Farm* in 2020 and Roger had to use his famous Glove of Love to help the magic happen between Lucky and an in-season sow. "We all need a helping hand from time to time!" says Roger. "Not to go into too many details, but a boar can't see what he's doing so it's all down to nature down there. And if that doesn't work, sometimes you have to guide them in the right direction." The boars certainly don't seem to have any complaints.

When Cannon Hall Farm first opened to the public in 1989, the Nicholsons bought a pot-bellied pig called Maggie and a pot-bellied boar called Denis. As Roger had never been called upon to don the Glove of Love for the two of them, he had assumed that they were not interested in breeding, only for Maggie to deliver a litter of teeny weeny black shiny piglets. "They were very bonnie, that's for sure and we kept them on the farm for the visitors to enjoy."

Like the pigs that his father used to keep back in the day, Roger keeps hybrid sows – Large White Cross Landrace varieties – and mates them with Hampshire boars. To avoid all of the sows having piglets at the same time, they are bred in sequences of threes and fours.

"This means that three or four sows will come into season at the same time and we can average out the size of the litters," Roger explains. "That way, if one has too many piglets, we can adopt them on to a sow who's had fewer. As long as you do it within a couple of days of birth, it works fine. You can see them smelling the piglet as if to say, 'I don't think that's mine', but they just start feeding and it's as simple as that."

Sows come into season for two days every three weeks and it's up to Roger and the farmers to make sure that the sows and boars are where they need to be to make sure the magic happens (with or without the help of Roger's Glove of Love). When the sows stand in a rock-solid position in front of the boar, that's his signal to move into action. "The longer that they hold their position, the more chance you have of a successful mating and a fruitful number of piglets. The boar knows when the sow is in season and if you ever put a boar with a sow that isn't ready, he just loses interest and will go and lie down – as if to say, 'What are you wasting my time for?'"

As well as being open to the public, Cannon Hall Farm is a working farm, and so when the piglets get to five weeks old, they leave their mums and go into a big barn with the other litters. One little piglet, though, recently had a reprieve. "We'd had a couple of boars that hadn't been fertile, so I kept back one of my crossbred pigs," says Roger. And the young boar's name? Lucky, of course! "He's fairly magnificent actually," says Roger.

17

Meerkat mayhem

"When Rob said to me, 'Let's get some meerkats', I thought what the heck do we want them for?" David remembers. "So it certainly wasn't my idea. But then, when I saw how the public made a beeline for them as they came into the farmyard, I thought, 'Fair-dos, Robert, it was a brilliant idea!' And they've turned out to be one of my favourites on the farm."

Meerkats also seem to be a documentary-makers favourite and in January 2009 the curious little animals became the face of the price comparison website Compare the Market, as cartoon characters that inexplicably spoke with east European accents. In real life, meerkats are found in southern Africa and are famed for their ability to stand on their back legs on lookout. They're a type of mongoose with a long, pointed snout, large eyes, long legs, a thin tapering tale and a taupe-coloured coat.

"In 2017, an exotic animal breeder that we knew called us to say he had a breeding pair of meerkats that we might like for the farm," says Robert. "As we had our zoo licence by then, we were able to keep more unusual animals than the usual farmyard varieties, so we decided to do some research on whether we could look after them. We have a team of builders at Cannon Hall Farm, so we Googled the ideal habitat for meerkats and got the builders to create a potential design. They did a brilliant job and came up with a really smart enclosure for them with tunnels and passageways. It's very cool."

After welcoming the new meerkat lovebirds to their swanky new enclosure, it wasn't long before they started to multiply. In the wild, litters of three to seven pups are born, so the farmers were delighted when their two gave birth to six babies.

"The dominant female in the group is Georgina," says farmer Kate. "She always gets the first dibs at any food and the comfiest spot when she wants to go to sleep. While all the others have to take turns going on lookout duty and making sure there are no predators, she always manages to avoid lookout duty. Ringo is known as the dominant male in the group, but he's got a very relaxed attitude to life and he doesn't do much to discipline the children – that's all Georgina's work."

"When you go into the pen to feed them, they're all over you, because they will always get to the highest spot they can find," says David. "But although they look really cute, they've got really sharp teeth, so we don't cuddle or stroke them because they'd bite you as soon as they look at you. They smell pretty bad, too!"

At Cannon Hall Farm, the farmers try hard to replicate the diets animals eat in the wild, and meerkats are no exception. "We give them mealworms, morio worms, locusts, crickets, bits of chicken and hardboiled eggs," says Kate. "They also love grapes, blueberries, sweet potatoes, greens and pumpkins. It's easy to overfeed them, though, so we make sure to give them a nice balanced diet to keep them slim."

Meerkats are very territorial – hence the constant looking out for invaders and predators. "When Robert walks past the enclosure with his dog, they all run up to the glass, arch their backs and start hissing at him," says David. "They're like little lions."

18

Sheep!

Where would a farm be without sheep? And where would Cannon Hall Farm be without the likes of Grizzly Bear, Tiny Lamb, Arnie and the gang?

Sheep might not be the most exciting animals to look at on the farm, but they play a vitally important role, as Robert explains: "Our flock is split into a number of rare breeds that we breed in order to help increase their number and sell on as breeding stock. Plus we have around 300 sheep which are commercially farmed for the farm shop."

They may all have four legs and a woolly coat, but the different breeds of sheep are as diverse as the places from where they originate. And in true Cannon Hall Farm fashion, there are lots of different personalities among them...

Take Arnie the Jacob ram. Not only has he got four impressive looking horns (hence his Terminator-inspired name), he is also one of the most handsome rams on the farm – and he seems to know he's a bit of a looker. "When we went to buy him in autumn 2018, we wanted a strong, good-looking ram that was going to attract the ladies," says Robert. "At that stage, we had 30 Jacob ewes and when we saw Arnie we knew he was an absolute ladykiller."

Having to fork out much more money on Arnie than they orig-inally intended, Roger wasn't so sold on their new purchase, but Robert was adamant the ram was a good buy. "He was definitely the best ram there on the day," says Robert, "and there were three different parties after him, so we had to up our budget a bit."

Back at the farm, the Nicholsons wanted to make sure that Arnie was going to deliver the goods and Yorkshire Vet Julian Norton was the man for the job. Armed with a strange vibrating gizmo, a tape measure and a microscope, Julian set to work checking Arnie's vitals were in working order. Meanwhile, Roger got into position near Arnie's nethers to capture the fruit of his loins, "I wouldn't get too close," Julian warned.

It was another moment of television gold as, on closer inspection, it was established that Arnie was indeed a great investment, and the following morning he was out in the fields with the Jacob ewes, earning his keep. "It was a big relief for us," said Robert. "When you've spent a significant amount of money on an animal, it's nice to know you've done the right thing."

Later in the year, Arnie got another chance to bond with the ladies and, to make doubly sure that the ewes were ready for him, Robert and David borrowed a couple of other rams from a nearby farm to be used as "teaser tups". "The presence of a teaser tup helps bring the ewes into season at the same time," Robert explains. "This makes for a more compact, cost-effective lambing season. In a sense, they are the warm-up for the main act."

Once the teaser tups have worked their magic, the Jacob ewes' woolly nether regions are carefully shorn so that Arnie can effec-tively do his job.

To keep Arnie's own ladykiller looks in order, David has the task of shearing the beast, who tops the scales at more than a 100 kilos. Holding the mighty Arnie in position while keeping his two pairs of horns in check is no mean feat. "He's called Arnie for a reason – he's strong, powerful and he's a terminator, but shearing has to be done," says David. "At one time, the wool cheque, the money received for fleeces, would pay the rent on your land. Now it sometimes costs more to pay a contractor than you would have paid for your sheep, so it's very much a welfare issue now."

It's an annual job and once Arnie has been fleeced he returns to the fields with a spring in his step. "We don't just shear them for their comfort," David explains. "The wool gets very heavy on their backs and a sheep will die if it can't roll over, so we need to ensure there's no danger of this happening."

Needing not just one shear per year, but two, the Swiss Valais sheep require more maintenance than other varieties, but the breed has become a real farm favourite – with Grizzly Bear at the top of the charts. "If a fine artist painted a Swiss Valais ram, he couldn't be more beautiful than our Bear," says Robert. "When we went to buy him, we knew he was just the perfect physical specimen. His horns, conformation and markings were all first-class and we knew he was going to add real quality to the bloodline we have on the farm."

Robert and David travelled to the far north of Scotland in order to buy Grizzly Bear from a Swiss Valais breeder. They had a maximum of £2,000 to spend, thinking that would be enough to secure the handsome tup (ram), but the breeder was expecting a lot more. "In the end," says David, "we had to say, 'Look, we've got £2,000 in our pocket and not a penny more, but we really, really want him.' Luckily for us, he agreed."

With his soft fluffy white fleece, little black feet, black face and bright shiny eyes, he looks more like a teddy bear than a grizzly bear. He even has the little black shin pads and a black pattern on his bottom that typifies the breed. All in all, he's a fine example of the Swiss Valais, which, as their name suggests, originates from mountains in Switzerland.

He certainly looks the part, but Grizzly Bear's first encounter with the Swiss Valais ewes was less than dynamic. Firstly, Stanley the goat had to take the lead as the breed is notoriously stubborn about being moved from place to place – and then Grizzly seemed to have more of a taste for Yorkshire grass than a taste for the ladies and he settled down for a good graze. Given time, though, he soon settled in and fulfilled all the requirements.

"We really wanted to do our bit for the breed, as the Swiss Valais has a very shallow gene pool," says Robert. "Sheep naturally breed in the spring of every year, but Swiss Valais can be bred all year round, so that's why we had autumn lambing in 2020. We're pleased to say that they are out of the 'rare' breeds bracket now, but they are still prized as breeding stock."

From Switzerland to Holland. The Cannon Hall Farm's collection of Dutch Spotted sheep is headed up by Vincent the ram, who was bought at auction in October 2019. In series three of *Springtime on the Farm*, the cameras followed Robert and David as, armed with a budget of £1,000, they set off to purchase a ram from this relatively new breed. David had never bid on anything except a marrow at the harvest auction, so it was up to Robert to use his trademark wink at the auctioneer to secure the sale.

"There were only seven Dutch Spotted sheep in the sale and I ended up buying three of them," Robert says. "But I knew that dad would really like them. We have loads of sheep that reflect farming's history, but I think Dutch Spotted Sheep are a breed we'll see much more of in the future, so I think it was £1,000 well spent. When you buy the best possible stock you can, you end up with more and more premium animals as you breed each new generation."

Ewes are pregnant for between 145 and 151 days, and generally give birth to single, twin or triplet lambs. Robert and David have been birthing lambs for as long as they can remember and know all of the potential problems they may encounter. "Dad had taught us everything we know, and we can conquer most problems," says David. "We've had nights where we have delivered 30 lambs, but many lambs, like the Soays, will do it all by themselves."

Soay Sheep are one of the hardiest of the rare breeds at Cannon Hall Farm. They are small, fine-boned brown sheep and come from a small island off the coast of Skye, in the Inner Hebrides of Scotland. There they feed on seaweed, shed their wool once a year and lamb by themselves. "They're just the perfect independent sheep," says Robert. "They are only tiny, but they are fabulous little things that know how to suckle and they're great mums. We don't feed them seaweed, though, just good old Yorkshire hay that we grow at Cannon Hall Farm."

Herdwicks are another hardy sheep variety now living at the farm.

"We got them because they are synonymous with the Lake District and it's an area that we love," says Robert. "We got our first batch from a wonderful hill farm up there. The farmer had two of the most incredible sheep dogs I've ever seen and they were able to round up his entire flock and keep them cornered while the farmer picked out 10 ewes we liked the look of. It was the most impressive display of shepherding I'd ever seen. It certainly got me thinking about having a sheep dog of my own one day…"

When it comes to crowd pleasers, sheep don't get more entertaining than the Zwartbles. The friendly, elegant Dutch variety, known for their long back faces striped with white, are easily tamed and very athletic, and so are the perfect candidate for sheep racing. "We saw that other open farms were doing sheep racing and it seemed a fun thing to do," says Robert.

Sheep are let loose on a short track in the racing arena with the promise of a tasty treat at the end of it. But don't worry they aren't starved to make them run faster. "We give them plenty to eat, the cereal we give them is just a Brucie Bonus on top of their usual grass."

With names like Calamity Carol, Dashing Derek, Rum Bah-Bah, Mutton Chop and Ed Shear-an, the racing ewes are a popular attraction at Cannon Hall Farm. And it's a two-way thing, as the sheep also seem to love it. "We've found that the fact that they've been raced makes them a lot quieter and a lot better at motherhood," says David. "Racing them is good exercise and enrichment, and they get a food reward at the end of it. We only race them for a short while when they are young, then they get to join the breeding flock."

With 11 different breeds of sheep on the farm, for a non-farming outsider it seems nigh on impossible to be able to keep track of which lambs belong to which ewes, which ones are pregnant and so on. In a modern business like Cannon Hall Farm, you might expect state-of-the-art systems for rearing sheep, but low-tech practices like numbering lambs with graffiti spray in different colours for singles, twins and triplets does just the job.

"It seems basic, but this way, if a sheep gets lost, you can easily reunite the whole family," says Dave. "We also mark the tups' chests with a crayon block so that when they mate the colour is transferred on to the ewes' backs, so we know which ones are likely to be pregnant."

It may be a basic science, but similar systems have been used for hundreds of years. "Keeping sheep is hard work," says David, "but if you get the lambs to the right birth weight it makes shepherding a lot easier. If you get really weedy lambs, they often have to be handfed, or if a lamb is too big the ewe may need a caesarean. It's hard to overfeed twins in the womb, but of course there are exceptions."

This was demonstrated by Tiny Lamb, the twin brother to Giant Lamb. "When it was born, it could fit into the palm of my hand," says Robert. "I thought, 'Oh 'eck, this is going to take some rearing.' I put him under a lamp and fed him and luckily he had the most determination of any lamb I've ever known."

Making the decision to take Giant Lamb away from his mum and hand-rear her was the only way to guarantee the survival of his tiny twin. "Tiny could hardly even reach her teat and he stood there on his trembling legs and then started sucking like mad. As he's a crossbreed from our commercial flock, we wouldn't breed with him – only purebred rams are used – but he showed such fighting qualities that we are keeping him on the farm."

He may be hard to recognise now, as Tiny Lamb is now Fully-Grown Lamb, but he's just as cherished for proving his determination to stick around.

Hand-rearing a lamb is a big commitment, as there is a lot of work involved, as was demonstrated when twin Zwartble lambs were discovered out in the paddock next to their recently deceased mum. "We decided to call the twins Annie and Oliver and they were hand reared with our other orphans," explains farmer Kate. "But if you're hand feeding one animal, you may as well do 10 or 20, so we get stuck in and make the best of it. There's so much satisfaction seeing lambs being reared ready to go outside. It's a lot of work, but a lot of pleasure as well."

Pip, Pip hooray

S heep dogs are as much a part of a farm as sheep and cattle, so visitors to Cannon Hall Farm are often surprised that the only sheep dog that has been around for many years is the one that appears on the sign at the entry to the farm.

The dog featured there is Flossie, Roger's favourite border collie, who lived on the farm in the 1970s. Due to problems at birth, Flossie wasn't particularly effective when it came to rounding up sheep, but Roger kept her as a pet rather than a working dog. Since then, the shepherding has been handled by the farmers – with a little help from Stanley the goat, who likes to get involved when his organisational skills are required.

Robert, though, had long harboured dreams of having a sheep dog help them with their growing flocks and in 2020 finally took the plunge. Enter Pip the sheep dog.

An episode of *Friday Night on the Farm* followed Robert and David as they took a road trip to Northumberland to meet renowned sheep dog trainer Emma Gray.

Robert had set his heart on getting a bitch, because, he says, "I think they have a kinder temperament."

"But the first dog that Emma showed us, Carla, was like a

Ferrari," says David. "She was raring to go. To be honest, I don't think Robert would have been able to handle her."

Robert agreed. "She was a beautiful dog, there's no doubt about it," he says. "I'm sure she's a future trialling champion, but she wasn't affectionate enough and I think she needed someone who could really boss her and knew what they were doing. It didn't seem fair on her because she has so much potential."

Then they were shown Pip. "If Carla was the Ferrari, then Pip's the family saloon car," says David. "She bonded with us really quickly and there was just something about her that felt right. She's calm and steady and not hot-headed."

Taking the border collie back to the farm in Yorkshire, Robert was initially a bit worried that she might not take to him as she had been trained by a woman, but it was love at first sight for Pip and she settled into Cannon Hall Farm straight away. "To train a good sheep dog they need to love you and she does that and she follows me, she wants to please and I want to do her justice because taking a working dog on is a big responsibility."

A sheep dog doesn't fully mature until it is four years old, so it needs time to enjoy being a puppy and take in all the basics before it is allowed to work with sheep.

Getting Pip at two-and-a-half years old meant that Roger had plenty of time to bond with her and enjoy having her around before the really hard work was to begin.

"Emma the trainer recommended that Pip doesn't live in the family house as she's a working dog," says Robert, "And although that might sound cruel, rest assured we have some lovely kennels and during the day she lives next door to Doris, my wife Julie's border terrier, who is being treated for bladder cancer. She was diagnosed three years ago and given just a few months to live, but thanks to wonderful veterinary care she is still going strong."

Pip might not have the power of her Ferrari-like sibling Carla, but she was raring to go when Robert took her on her first grand tour of the farm. She had never seen pigs before, so she was particularly interested in the piglets, and she was especially excited when she saw the lambs in the Roundhouse. "This was good news for me," says Robert, "Because sheep need to be the focus of her attention – the epicentre of her life."

Training a sheep dog takes commitment. "I've been taking her training very steadily," says Robert. "I do about 20 minutes' worth every day, practising 'come by', which means left, 'away', which means right, and 'lie down'. Pip is very keen, but our sheep aren't used to having a dog boss them around, so it's been a bit of a steep learning curve. It's a dog's instinct to bring the sheep back to you, but the sheep didn't want to play ball and stamped their feet at her. So it's been a case of gradually driving the sheep around the edge of the field so that they could get used to her.

"Pip seems very happy, though and that's the most important thing."

20

Fun with the ferrets

Not many people can boast that their uncle sponsored the World Champion Ferret Legger, but then again, most people aren't like the Nicholson brothers...

The sport of ferret legging was popular for many years, particularly in Yorkshire mining communities. It saw brave, some would say foolhardy, competitors place a ferret down a leg of their trousers which had been tied at the ankle so that the ferret couldn't escape and then timing how long the competitor could stand the pain of being scratched and bitten as the poor ferret tried to find its way out. And, by the way – no underpants were allowed.

With four long canine teeth to kill prey, and rows of razor-sharp incisors, ferrets can inflict serious harm on a predator, and being shoved down a trouser leg must have scared the bejeezus out of the unwitting creatures.

Thankfully, it's a sport that's died out, but the Nicholson boys remember how much they used to like ferrets when they were younger and now, having the space for them, they wanted to find homes for them on the farm. "Ferret legging is definitely a thing of the past and it's staying there," says David. "It used to be a test of manliness, but when they bite your manliness it's not a good test. Ferret racing, on the other hand, is something that everyone can enjoy! We first got ferrets on the farm about eight years ago. For some reason, my dad is scared to death of them – and he's not usually scared of anything! One day he came running down the yard saying, 'David! Come quick, come quick!' and I thought something awful had happened, but it was just a ferret strolling down the path."

"Our original set of ferrets were a real mixed bag," says Robert. Small white ones, a large black one that looked like a wild polecat, and a really fat one called Roly (short for Roly Poly). He was too fat to race because he couldn't fit along the perspex tubes."

Ferrets spend between 14 and 16 hours asleep every day, but need to have at least an hour of daily exercise to keep them mentally and physically healthy. "Racing enriches ferrets' lives," says Robert. "It gets them out, they enjoy what they are doing and people enjoy watching them doing it too."

It's unlikely it will ever be an Olympic Sport, but ferret racing at Cannon Hall Farm certainly gets the pulse racing – for both the ferrets and spectators alike. The ferrets start off in red, yellow, blue and green colour-coded wire boxes and are encouraged to run along a track of perspex tubes to reach a tasty reward at the end. Spectators can bet on the winner.

As you would expect at Cannon Hall Farm, the racing ferrets are suitably named, so there's Stoatally Different, Ron Weasley, Bryan Ferret and more…

"In Japan, ferret racing is taken quite seriously and the tracks are a lot longer than ours," says David. "But we just do it as a bit of fun. The wire boxes in between each section of perspex tube are wide enough for the ferrets to turn around and go back the other way if they feel so inclined, and it's always a laugh when that happens.

They might be leading the race, then they turn around at the last junction and head back."

The Cannon Hall Farm ferrets haven't as yet been featured on any of the Nicholsons' television shows, but they have appeared in a Christmas episode of *The Yorkshire Vet*. In a tribute to the carol, The 12 Days of Christmas, instead of "six geese a-laying", it was "six ferret castrations", as Peter Wright and Julian Norton went through their Christmas list of jobs that needed doing. Luckily for the ferrets, on this occasion two of them were spared, as they were still a bit too young for the chop. It would need to happen at some stage, though, as males need to be castrated to calm their aggressive behaviour and females can become infected if they are not mated or bred, so they are spayed or given contraception.

"Ferrets really are easy to look after," says Robert "And they make great pets. They might not be Dad's number one choice, but then keeping a bull like Jeremy might be thought of as a strange choice. It's all horses for courses, isn't it?"

21

Into the woods

Rocky the deer, who was rescued and rehomed at Cannon Hall Farm in the spring of 2020, holds a special place in Robert and David's hearts.

"We had a call from the local police saying that they needed our help with a deer that had been hit by a car on the M1," says David. "We met them at a slip road near Junction 27 and they escorted us up the wrong side of the motorway to the injured animal. The traffic was backed up for miles and there was blood all over the road. Rob and I didn't think the animal stood a chance."

"When we got there the deer was unconscious and bleeding from his mouth," says Robert. "We carefully lifted him into our trailer and he just lay there not moving. I thought maybe he had massive internal injuries and so I was fully expecting him to be dead by the time we got him back to the farm."

Although the deer was looking very shell shocked and had clearly been injured in several places, he actually managed to stand up when the RSPCA officer arrived to check him over. Knowing that Robert and David care for animals for a living, the RSPCA officer was happy to let them take over the deer's care and release him back into the wild when it was possible. After all, the boys were fortunate to have access to their very own wood on the Cannon Hall Farm estate.

As the accident happened during the first lockdown in the spring of 2020, the roads had been much quieter, which is probably why the deer had ventured onto a main road in the first place. But what with it being lockdown, it also meant that accessing a vet wasn't as straightforward as normal.

"We Skyped Matt from Donaldson's Vets," says Robert, "and he suggested that we gave the deer some Metacam, which is a pain relief and anti-inflammatory drug, and we made it comfortable for the night in the back of a trailer with plenty of straw and a bucket of water."

The following morning, when Matt stopped by the farm on the way to work, he was amazed to see how sprightly the little deer was looking. He suggested he could be named Rocky because of all the cuts and bruises on his body, and said the deer should be released back into the wild to give him the best chance of recovery. Robert, David and Matt then drove the trailer to the Deffer Woods at the edge of their land and unlatched the trailer door.

It took a while for the young stag to make a break for freedom, but he sprang off to the woods looking revitalised and full of the joys of spring.

"The best thing about releasing him was when he turned towards us just before he disappeared from view," says Robert. "It was as if he was saying 'Thanks, lads' and I wanted to tell him to have his best life and that I hoped we'd see him again in the future."

And so, happily, that's exactly what happened. "Around two months later I went back to the place where we had released him and I saw him!" says David. "We see deer in the woods a couple of times a year and I feel sure it was him. He was the right age and the right size and he looked so well."

"He's part of the farm and I feel it was our good deed for the day when we rescued him. Even though the motorists that had been disrupted probably wouldn't agree..."

22

The farm shop

The farm shop has come a long way since the days of selling his 'n' hers cow mugs and pop-a-point pencils. The award-winning Cannon Hall Farm shop is officially one of the best farm shops in the UK and one quick sweep around its colourful, flavour-packed aisles and it's easy to see why. Breathing in the aroma of their beautifully buttery Eccles cakes is enough to put anyone's diet on hold.

From the days of being part of the Cannon Hall Farm tea room and a rather lacklustre souvenir outlet, the farm shop is now a thriving business that features the very best local produce and, of course, meat from the Nicholsons' farm.

With manager Caroline Glover at the helm, butcher Alan Asquith curating the meat counter, Liz Ball (known as Barnsley's Mary Berry) heading up the bakery, and chefs Mark England and Tim Bilton rustling up the ready meals, it's a foodie's dream to browse and shop there. "The number of regulars we have is a real testament to how good our produce is," says Richard, who originally came up with the idea of having a farm shop to sell Cannon Hall Farm meat. "We have one customer that regularly travels down from the Highlands with three huge storage boxes and buys £1,000s worth of meat at a time. As Scotland is famous for the quality of its meat, it's proof of how good ours is!"

With the wealth of local produce on offer, from Yorkshire Parkin gin to an entire Sunday roast ready meal served in a Yorkshire Pudding, the farm shop ticks all the boxes for locally sourced products and low food miles for the farm's beef, lamb and pork. "Provenance, traceability and the farm-to-fork relationship is very important to us," says Caroline, who is also one of the main buyers for the shop. "I have a local, old-fashioned milk supplier who milks everything herself and brings it to us in traditional glass bottles with silver tops. It's the nicest milk ever.

"I want to keep it as much of a traditional farm shop as possible, but at the same time keep it different to others. That means along with the daily fresh fruit and veg that you'd expect to see and a great selection of local cheeses, honey, jams, chutneys and meats, you'll also find things like Italian, Indian and Chinese specialities, hand-made Belgium chocolates, artisan kitchenware and restaurant-quality ready meals."

The farm shop is a real success story, as its impressive award tally proves (Taste of Yorkshire 2017, Best UK Farm Shop 2018 and Farm Shop and Deli Awards in 2019), and 2020, despite all the challenges it posed, saw it go from strength to strength. "We had all sorts of plans for the shop in 2020," says Caroline, "but we had to put them on hold and concentrate on fulfilling orders, which went through the roof during the two lockdown periods. Our staff were working non-stop at 100 per cent speed just to keep up with supply and demand."

With less access to normal suppliers, people living in the vicinity of Cannon Hall Farm discovered the joys of the farm shop. It certainly doesn't look like your average supermarket with its seriously impressive range of fresh goods and groceries. Caroline says: "Once people tested the quality we offer, discovered the provenance of everything and realised they really don't pay that much more than they would at a supermarket for much better-quality food, it opened people's eyes to farm shops."

You might not make a special visit to the farm shop to stock up on loo roll and toilet cleaner, but if you want to try something new, this is the place. How about a glass of rhubarb and custard beer, organic eggs from Pingle Nook farm, or some local Barncliffe brie? "I'm always trying to taste new things for the shop and if customers like it they'll keep coming back for more," says Caroline, whose all time-favourite product in the farm shop is Cynthia's scones. "If a product doesn't stick, we know it's not quite right for us."

Gone are the original Harrods-inspired gold and green carrier bags. Now it's all about the smart black and white packaging featuring Roger and Flossie the dog. Slowly but surely, the farm shop has built up from being a very basic farm shop selling a few bits and bobs, to the success it is today. "It's always surprising when I look back," says Richard. "When we combined the original deli and the farm shop, it was the catalyst for it becoming a better shopping experience."

Expanding the numbers of staff has been a gradual process. "You can't get by without having the right people working with you," says Richard. "For example, Mike West, the person who creates all of our food labels, is a really clever fella – a doctor in chemistry, no less, and he's a real detail person – and Robert, David and I really need that detail!"

It's the attention to detail that makes the farm shop such a great shopping experience. And that's why the customers aren't just the visitors to the farm that remember they need a pint of milk at the end of their trip to see the animals. The majority of shoppers are regulars who often insist on only being served by Alan the butcher and know which of the 20 flavours of sausage is their favourite.

"A lot of farm shop owners talk the talk, saying they supply locally sourced goods, but there are a few that don't walk the walk," says Richard. "The vast majority of what you'll find on our meat counter comes from our own farm and the rest is extremely local. But what's even more important to me is that the shop is a celebration of good things to eat – wherever the produce comes from, we want it to be the best of everything. That means we will always stock the best quality seasonal and local food, but we also sell year-round favourites. After all, if we only sold seasonal food, you'd soon get bored with parsnip and turnip soup!"

23

Meet the team

Introducing some of the talented staff that make Cannon Hall Farm so special...

Carol Rayner, General Manager
With her legendary laugh and larger-than-life personality, Carol is almost as much a fixture of Cannon Hall Farm as the Nicholsons themselves.

Having previously worked for British Telecom for 24 years, Carol joined the team as General Manager in 2002, and is responsible for staffing and all aspects of sales. Back at the start, it was, she says, "a little job". Now, of course, it's a much bigger role, with Carol heading up a 250-strong team.

She works alongside Robert's daughter Katie, who is the HR Manager. "We spin off each other. Katie has been and done all the qualifications and I've got the experience – we're a good team."

Zipping around from department to department every day, Carol prides herself in knowing every aspect of everyone's job, from working the coffee machines to cleaning the toilets. She hasn't yet worked with the animals or helped birth a llama, but never say never.

"I've been at Cannon Hall Farm for 18 years and although some days are incredibly hard – I have been known to be dealing with staff issues at three o'clock in the morning – I still leave each day knowing that the family think an awful lot of me. I will always do the best possible job I can for them.

"I love the work ethic here and the fact that, say, if the staff weren't coping in the restaurant, everyone – including Robert and David – will get their pinnies on and clear tables. The Nicholsons are very proud of their business. And rightly so."

Nicola Hyde, Marketing Manager

"My job interview consisted of having a full English breakfast with the family, then sitting in the soft play area with my son and Richard, while we played with floppy lego bricks together and discussed the role."

As a marketing consultant, Nicola was initially introduced to the Nicholsons at Cannon Hall Farm when she was involved with the rebranding of the White Bull Restaurant. But it soon became clear she had big ideas and she was offered a full-time job.

The former award-winning journalist now has a varied role which involves coming up with event concepts, such as the annual Pumpkin Festival, Easter Egg Hunt and Santa Experience. She liaises with the press, setting up interviews and helping out with publicity for the television shows; coordinates the social media videos; runs the Facebook Supporter's Network; works on the Cannon Hall Farm merchandising; and heads up the animal adoptions.

"What I love about working at Cannon Hall Farm is that the family is fearless and no idea is ever too crazy. For example, I might say, 'I want the joiners to build a three-room elves' workshop that our kids can walk through and then make a teddy bear,' and they always say, 'Right! Let's do it!' Nothing is ever a no. It just goes to show the extremely resilient nature the whole family has. Like many people at the farm, I'm pretty much given free rein to do my role, and that's the ethos across the farm."

With Nicola's background in creating a brilliant community website, broadcasting live from celebrity events and filming and editing videos, she has been instrumental in the Nicholsons' social media success. When she first joined the team, the boys were doing bits and bobs on Facebook, but she saw the potential of sharing what goes on behind the scenes. "I really wanted to emphasise the family nature of the farm and promote the animals' fantastic personalities. Rob started his regular live broadcasts and that's what brought us to the attention of Channel 5."

Alan Asquith, Butcher

He's known as the "housewives' favourite" at Cannon Hall Farm and while Alan is very modest and laughs off any kind of heart-throb butcher status, his dedication to both his customers and his profession is plain for all to see.

Having worked for Roger's sister Shirley for nearly 30 years, Alan joined the team at Cannon Hall Farm when Shirley's farm closed in 1995, and many of his customers followed him to his new location.

He has seen a lot of changes on the farm. "When the family built the café, I said, 'I bet that's not going to be big enough soon,' and I was right – within a year they had to extend it. And at one time we used to have just one counter and a butcher's block, now we have a walk-in freezer and there can be 10 of us behind the counter."

Still priding himself on the produce on sale, from choice cuts of beef to black pudding and potted meats, Alan says that even the famous Barnsley chop has changed over the years. "It used to be a long lamb chop from a Southdown sheep and it weighed one pound and six ounces. The Prince of Wales was presented with one when he came to open Barnsley Town Hall in 1933. Now most people think of the Barnsley chop as a butterflied double chop."

Alan's attention to detail and his incredible memory is one of the reasons why loyal customers keep coming back to the farm shop. "It's quite flattering, I suppose, and at one time I used to be able to remember what my regulars bought at Christmas from one year to the next. But I'm knocking on a bit now."

As the longest serving butcher at the Farm Shop, Alan is part of the furniture and will be for many years to come. "The best thing about working at Cannon Hall Farm is the buzz and the ambiance – just everything, really. I love meeting the customers and making sure they're happy. I get a kick out of knowing I've done a good job."

 **John Hopkinson, Health, Safety and Compliance Manager**

"Robert was in the first class I taught on my very first day of teaching at Barnsley College in 1985, and here I am, over 35 years later, still working with him."

John taught agriculture lessons to both Robert and David and was instantly impressed by the pair of them. "Robert is excellent on the theory side, while, on the practical side, David is one of the best students I ever had." Robert and David joke that between them they make one decent farmer!

Fast forward 35 years and Robert got in touch with John again. By that stage, John was working in farming consultancy and indulging his lifelong love of music, managing several bands and teaching music lessons. "When Robert offered me my job, he said,

'I want you to do all the things that I don't like doing' – meaning all the health and safety stuff. I knew exactly what he meant because, to be completely honest, if anyone had said 'health and safety' to me when I was younger, I would have run a mile to avoid doing it!"

John has the vital role of keeping staff and visitors safe and healthy while they are at the farm, making sure that all risks have been assessed and all insurance and red tape is adhered to. Each and every aspect of the farm, from the hand-washing facilities to the fencing around the animals, has to be painstakingly checked and double-checked, so that everything operates like clockwork.

"It's unbelievably different to when I first stepped on the farm. Back then, when Robert and David were students of mine, Cannon Hall Farm hadn't diversified and opened to the public, and it wasn't far from being threatened with bankruptcy and closure. To see how far Robert and the family have come is quite remarkable, and they have done a brilliant job in assembling such a talented team – and that didn't happen by accident."

Helen Mallinson, Office Manager

Having worked at Cannon Hall Farm for more than 26 years, Helen is officially the longest serving member of staff outside of the Nicholson family. Starting out as "chief rabbit stroker", Helen came to the farm for work experience in 1994 when she was 15 years old, then a year later started working in the gift shop while she was studying for her A levels. After completing her accountancy degree at Huddersfield University, it was back to Cannon Hall Farm, this time with a full-time job working in the office alongside Cynthia and Robert's wife, Julie.

"At that stage, I was doing a bit of everything because it was still quite a small business, so I was taking school bookings, paying suppliers' bills, invoicing, all sorts," says Helen. As well as spending five days a week in the office, for 15 years Helen also worked in the farm shop on Saturdays. Over the years she has seen the business grow from a small open farm to the incredible success it enjoys today. As Cannon Hall Farm's Office Manager and book-keeper, Helen still works alongside Julie and the human resources, marketing and sales teams.

"The great thing about this job is it's a totally unique place to work and you never know what a day's going to bring. I just love it here. I'm super-proud of what the Nicholsons have achieved because they have put so much hard work into the business. Cynthia is my best pal and my mentor. We spent so many years together and she taught me so much; she is so clued up about the business. I have so much love and respect for her."

Linda Wetherill, Farm Manager

"When I was offered my job at Cannon Hall Farm in 2005, they said, 'Welcome to the family' and there was never a truer sentence spoken. Hand on heart, the Nicholsons are the nicest employers you could wish to have."

Former police officer Linda Weatherill began work as a part-time tour guide after she'd had children, but since then her role has grown at the same pace as Cannon Hall Farm. "My official title was 'Tour Guide and Various' and the 'Various' bit is still somewhere in my job description and it often comes into good use!"

As the farm manager, Linda is in charge of staffing the farm side of Cannon Hall Farm, everything from organising staff to mucking out the large animals, small animals and reptiles, to manning the car parks, the farm entrance and the tourist attractions such as the tractor rides and the sheep racing circuit. She also runs the work experience programme with schools, colleges and universities, which is where the majority of her staff first started out.

"The Nicholsons are very good at just letting people use their initiative and get stuck into things. The tour guiding side of my job became busier and busier and I just fell into taking over running that and various other positions on the farm. I have knowledge that I came here with and the rest of it I've learned as I've gone along."

Originally there were far fewer visitors in wintertime, so Linda began working in the delicatessen and the gift shop, but as visitor numbers increased, multi-tasking and performing a number of different jobs became a thing of the past. Now each member of staff has a vitally important part to play and with her skills as a super-organiser, Linda will help find the right person for the job.

"The staffing itself is no mean task and there's quite a lot involved, but I have a fabulous team and they work their socks off. I have to say the Nicholsons are a wonderful family. What you see is what you get. Hard working, down to earth and caring individuals. I could never imagine wanting to work anywhere else."

Liz Ball, Bakery Manager

"Fifteen years ago two friends asked me if I'd help out with the Christmas rush, and I said 'I'll do three weeks.' I'm still here!"

Like so many of the dedicated Cannon Hall Farm staff, Liz never imagined she'd be part of the journey from a small farm to the success story it is today. Initially she worked in the delicatessen bake shop, making "a few pies" and now she heads up a team of 10 that every day create around a dozen varieties of pies, 14 different loaf cakes, around 15 types of buns and all sorts of tarts including all the favourites.

It was perhaps inevitable that Liz would end up being Barnsley's answer to Mary Berry. Coming from a family of 10 children, where everyone joined in with the cooking, Liz and her husband owned a restaurant for 10 years and a hotel for five years. So Liz is used to running a tight ship. "I've only been the manager here for five years but I'm really pleased with what we have achieved because the turnover has more than doubled. It's probably because I'm so bossy and I don't miss a trick! I can watch everyone working at the same time as doing my own job and I even keep an ear out for the machines around the corner to check they don't stop. I have a great team here though, I'm very lucky."

Springtime on the Farm and *This Week on the Farm* fans will recognise Liz from when she judges Rob and Dave's baking competitions. "They're very competitive – Robert is anyway – I think David just goes with the flow to be honest! I know I should be tougher when I judge them, but it's hard – they're like family!"

Mike West, Assistant Office Manager

From slicing ham in the deli to working out how to run 28 till points on the farm, making sure that every chemical used on the farm is safe, that all the products are labelled correctly in the farm shop and the ticketing system works, Mike West has another of those jobs at Cannon Hall Farm that has evolved dramatically since he joined the team in 2005.

"Before coming to work for Cannon Hall Farm, I was a senior chemist at a company that produced chemicals for other companies

– things like the pigments in coca cola cans and the brightening agents for laundry powder, but we were fighting against competition in India and China who could make products more cheaply. In the end I decided I wanted to do something totally different."

Although Mike may have started out with a part-time job on the deli counter, his skills as a 'detail man' were soon discovered. "Cooking is like chemistry really. You put raw ingredients together, mess around with them and end up with a product you want. When we started producing our meat pies etc, we needed to create all of the correct labelling to adhere to all of the right standards and it became my job to ensure all the paperwork was in order. I then moved into the office to work with Helen and my role expanded."

When COVID-19 hit and all of the visitor tickets had to be pre-booked, that meant an overhaul of the system that, luckily, was just the job for Mike. "We had previously only pre-sold tickets for big events like Visit Santa and then suddenly everything had to go online. It was a big undertaking troubleshooting potential pitfalls.

"Mike forensically dissects our impulsive ideas and makes sense of them," says Richard. "He's such a detail man and would definitely be your 'phone a friend' on *Who Wants to be a Millionaire'*."

"I've been here for 15 years and as the farm has evolved gradually it's easy to think that nothing has really changed, although we all seem to be working harder! However if I look back and compare what it was like to when it started and now - it's wow!"

Tim Bilton, chef

"The first time I came to Cannon Hall Farm, I was completely taken aback by the passion and the care that the Nicholsons have for their animals. The way they raise them and their whole ethos of farming ticked the box about everything I was trying to achieve."

Back then, Tim had two very successful restaurants in Yorkshire and he would buy all of his lamb and beef direct from Cannon Hall Farm. Tragically, he developed a rare form of cancer and was forced to step back from the restaurants. Following extensive treatment and a long road to recovery, he felt fit enough to work again – at exactly the same time that Robert was looking for someone to add the wow factor to the food served at the Cannon Hall Farm restaurants.

Having a head chef who was a finalist in the BBC's *Great British Menu* and trained by some of the best in the business, including Raymond Blanc, was a real coup for the Nicholsons and also the perfect fit for Tim. Sadly, he had to take more time out when cancer hit again in March 2019, but since autumn 2020 he has been cooking up a storm in the farm shop. "I love Cannon Hall Farm and on a personal level I am completely indebted to how the whole Nicholson family have supported me – Robert, Richard, David and their spouses, Robert's daughter Katie in HR, and obviously the two chiefs, Cynthia and Roger. They have all been amazing to me."

Dale Lavender, farmer and film-maker

From car park attendant to wannabe Spielberg... Dale has come a long way in the 13 years he has worked at Cannon Hall Farm. You'll recognise him from his various films on the farm's website, Facebook Lives, Insta stories and YouTube, where he checks in with various farm animals and keeps us up to date with all the farm goings on.

"When I was at college, I started working at the farm part-time – that's when I was manning the car park. Then I got a job as a general farm worker and started mucking out the pigs and feeding the animals, and it's just gone on from there."

As the variety of breeds increased at Cannon Hall Farm, Dale learned more about each animal's nutritional, handling and training needs, working closely with Roger, Robert and David. "When the farm started to build up its social media presence, we all started doing little selfie videos and it coincided with me getting a new digital camera. I started experimenting with filming and editing and watched internet videos about it and gradually did more and more films for social media."

In 2020, when lockdown hit, Dale's skills came to the fore. "I was out in a field worming some sheep when Dave told me he needed me to do the films for *Springtime on the Farm*. It was nerve-wracking in one way, but I thought I'll give it my best go and if my best wasn't good enough then I'll soon know one way or another!"

With lots of online communication with the television production team and tips on how to film various shots, Dale was soon creating great films that were used on the programme. "I always have imposter syndrome and think what I have made is rubbish, so there must be some good editing going on in the TV production company."

When he's not working behind the camera, Dale is happy to be back with the animals – especially Lottie the shire horse. "It's easy to take things for granted, but when we see how other people are around animals, I realise how lucky I am to have so many special moments with them, like when a new animal is born and survives against the odds.

"I don't know about being the next Spielberg, I'm just happy staying right where I am."

Darrell Mitchell, farmer

Want to know how long a porcupine is pregnant or where in the world Mangalitsa pigs originate? Farmer Darrell's the man to ask. You will often see him accompanying school groups around the farm, explaining all about the animals. He is such a natural that people have asked him if he was ever a teacher, but before he joined Cannon Hall Farm he was in a very different profession to both teaching and farming.

"I was a publican, but I always wanted to work with animals and I decided I wanted to have a complete change of life, so in 2014 I applied for a part-time job at the Nicholsons' farm."

Like so many members of staff, that part-time job soon turned into something more permanent, which came to involve training up other tour guides, getting hands-on with the animals and making social media videos with Dale.

"I go on the internet and research some fun facts about the animals at Cannon Hall Farm, then Dale and I record a little film with lots of information about the particular animal. Then I test viewers about what they've learned with three little questions at the end. We encourage children to message us their answers, draw pictures and come up with names for the animals to make it a bit more fun."

Darrell's personal favourites on the farm are Prince and Jeffrey the reindeer. "They are incredible how they have adapted to living in the Arctic, they are such well designed creatures." He also has a particular penchant for Spike the porcupine.

"Working with animals every day and caring for them is a million miles away from my old job but it's so rewarding. I really think agriculture and farming should be on the national curriculum because children get so much out of a visit on the farm. And I get so much job satisfaction seeing their reactions to the animals. It's just magic."

Kate Bodsworth, reptile expert

Farmer Kate first came to Cannon Hall Farm in 2011 as a 16-year-old work experience student when she was studying animal management at college. She loved it so much she continued to help out at the farm whenever she could at the weekends and during her holidays. She then went on to study animal behaviour and welfare at Lincoln University and, as a result, walked straight into a full-time job at Cannon Hall Farm.

"I always knew I wanted to work with animals, but I didn't know exactly what I wanted to do," she says. "I was always more focussed on mammals, but when I heard we were getting reptiles at the farm I put a lot of work into researching their needs."

Kate's fascination for exotic animals intensified through her travels to South and Central America where she first encountered leaf cutter ants in the wild. "I was part of a tour group and I started properly geeking out, telling everyone all the facts about these ants!"

Not surprisingly, you are most likely to find Kate in the Reptile House with Iggy, Rex and company, or hand-feeding the baby animals that need extra care during lambing. She's also one of the regulars on the farm's Facebook Live broadcasts and looks after the meerkats and Spike and Winnie the Porcupines.

Ruth Burgess, equine expert

"It never stops here!" says Ruth Burgess, who has worked at Cannon Hall Farm since 2017. Having worked at countless farms and taught farming at college, she is the resident equine expert and has trained the likes of (Awkward) Orchid, Lottie and Blossom, as well as working along-side many of the other farm favourites.

"I've never classed myself as a 'horsey' person," she says. "There's just something about the big-boned breeds. They're so majestic and they're part of our farming heritage, so I want to do my best for them."

Her best memory of being on the farm is seeing Orchid's foal Will being born. "I'd been watching Orchid for weeks and weeks as she got near to foaling and one night David rang me at home to say it was happening. I live about 20 minutes away, but that night I just flew here – how I didn't get a speeding ticket I don't know! It's been so special seeing Will grow up here. He's a stubborn little monkey, but he's so special."

An average day for Ruth involves feeding, mucking out and checking the health of the various animals; grooming, exercising and training the horses; engaging with the visitors; and keeping up with all the paperwork that is needed for Cannon Hall Farm to keep both traditional animals and those classed as zoo creatures. "Every day is different, depending on what needs to be done, but it's such a great place to work because you really feel like one of the family."

Poppy, Anita, Henry (Poppy's fiancé) and David

Left: Marshall with Primrose and Millie. Middle: Marshall with Clare. Right: Tom and Ka

Left: Katie, her husband Rob, baby Nelly and Dolly the sausage dog. Right: Julie with baby N

2 4

It's a family affair

The Nicholson story has always been about family working hard together to build a future and to be able to live their best lives. Several other members of the family live and work at Cannon Hall Farm and the family's other farm a few miles away, and share an incredible bond. Together they are part of the farm's past and its future.

Here, the brothers' partners and their children recall some of their favourite memories of life at Cannon Hall Farm.

Julie, Robert's wife

"Robert and I got married on Saturday 18 March 1989 and we opened the farm the following Friday. It was so exciting knowing that it was going to happen after so much planning. Back then I was working at Home Farm tea room, having previously worked for an animal feed company. In true Nicholson style, Cynthia and Roger offered me the chance to take it over, so I thought I might as well have a go. My grandmother was Italian and when you have grown up with Italian ancestry you show your love by feeding people, so we loved to cook and entertain and eat. I worked there for around 15 years and then we set up an office and Helen Mallinson and I taught ourselves how to do payroll and accounts. We had about 13 people on the books then – a bit different to now when we have around 300 staff when we're at our busiest."

Katie, Robert and Julie's daughter

"I remember when Mum, Helen and grandma Cynthia were working in the 'office' in our house and our black labrador Belle had a litter of puppies that we kept in a paddling pool in the office. The house was always full of animals – another time we had a poorly baby donkey by the radiator keeping warm. I used to have two goats and put a collar and lead on one of them – Christine – and lead her around."

Tom, Robert and Julie's son

"Growing up around so many animals and always having so much access to them is a lot of people's dream childhoods – the fact that you could just wander into the farm and entertain yourself with all sorts of things. And it wasn't even a normal farmyard because of the range of animals. For a long time, I wanted to be a vet, but I was a bit put off when I watched a baby calf being pulled out of a cow and I got a bit squeamish. After that I was going to be a small animal vet – until I realised that science wasn't my best subject!"

Poppy, David and Anita's daughter

"I'm told that when I was about two I was attacked by a cockerel and then I had a persistent cough and it was discovered I was allergic to horses, goats and other animals, so I almost feel the experience of growing up on a farm was wasted on me. Over the years I've got better with animals, though, and I love hand-feeding the newborns. When I was little, the best thing about living on the farm was having Katie and Tom – my two best friends – living right next door. We had free rein of the playground when the farm was closed. Our mums would just stand on the doorstep and shout us in when dinner was ready."

Marshall, Richard and Clare's son

"My first memories of the farm are sitting in my dad's tiny old office and finding pictures of the animals on his computer and printing them off for my scrapbooks. I also have really early memories of the play area, especially the tube maze, which was one of my favourites. I loved having the freedom to run around everywhere. It felt like I had a big playground all to myself."

Julie

"When Tom was about three and Katie was a year old, my parents decided to up sticks so that they could see more of Tom and Katie, because time was precious, so when a house in the courtyard became available they bought it. At the time I suppose we took it for granted, but it was wonderful never having to worry about childcare or having to leave the children with someone I didn't know. They've been there ever since. It's a bit like being in *The Waltons*!"

Anita, David's wife

"When they were tiny, Cynthia would look after Poppy, Tom and Katie in the office and she'd have a child on each arm of the chair and one on her knee. When she had to take a school booking, the children knew they had to keep very quiet. They were taught from the age of two that when the phone rang, the story ended as Granny had to work, but once the phone call was finished the *Benjamin Bunny* story could continue."

Poppy

"We didn't dare say anything when the phone rang! We were so well trained. We used to call Cynthia Granny Grumps, but we were never scared of her or anything like that – it was just her sense of humour. When we were a bit older, we used to make booby traps all over the office with clumps of elastic bands which we'd stretch over furniture. It used to drive Granny mad! I know it sounds silly, as we only lived two houses away, but I loved staying over there because Granny and Grandpa always had porridge every morning and Granny would always read me a book at bedtime. She was really good at reading stories."

Marshall

"Granny and Grandpa looked after me quite often when I was much smaller, and Granny had a little stash of games just for me in her drawer. I remember covering their fridge in Peppa Pig magnets – I don't think they minded, though. They both often played with me, but I don't think Grandpa liked losing – I remember once that he got mad when I beat him in a game of hoop toss. I was only five or six!"

Poppy

When I worked at the farm, Grandpa used to infuriate me when he insisted on helping me carry the animal feed up the hill. I used to say, 'It's because I'm a woman!' But he just wouldn't hear of it, he's such a gentleman."

Anita

"Cynthia is the most generous person you could ever meet – she's the kind of person that if she had ten pence she'd give three pence to each of her sons and leave herself short. She and Roger have instilled such a great work ethic into everyone around them. Money was tight for a lot of our lives, but Poppy has grown up knowing how important it is to work extremely hard and save the pennies – before she went to university, she saved £2,000 from working all summer."

Julie

"Roger and Cynthia have always worked so hard for whatever they have, and they've always said they would never make anyone do anything they wouldn't do themselves, and that's the same for all of us. They are both striving for the same thing – to pass something on to their family. Then our generation will pass it on to Tom, Katie, Poppy and Marshall, and hopefully they can continue that."

Clare, Richard's partner

"My background is in the travel industry, so completely different to Richard's, and although my friends say I went from being a city girl to a farm girl, I still do the same job, I just live on a farm, that's all. When I first visited Cannon Hall Farm, it was a decent-sized going concern, but it's grown massively. The whole family is lovely – they are such down to earth, humble people, particularly Roger and Cynthia, who still look like they are scratching their heads, not quite believing the success of it all."

Katie

"It's hard to live up to Grandma and Grandpa because they genuinely are the most amazing, loving, generous, funny people. They're like my friends. And the fact that my grandparents on my mum's side live in the courtyard is so wonderful. I'm still really close to them and I see them most days. It's wonderful for baby Nelly, too."

Tom

"As you get older, you grow more of an appreciation for your childhood, and I can look back with far more clarity on how wonderful it was to be on the farm and have my grandparents around me. My dad, uncles and Gran and Grandpa have been able to turn around a farm that wasn't an economic business prospect in any way into something that is now very successful. And although I live in London now, I talk to my dad nearly every day about the business – the merchandising and social media as well as the wider business strategy – and it's so rewarding to be involved."

Marshall

"I like to go out and check the sheep are all okay at home in Mill Farm for Grandpa most days, sometimes I take Riley our dog with me. I help Grandpa and also Rob and Dave feed the animals. I've helped sheep stuck in mud, bushes and even guided a couple that fell in the pond to safety! I've also done a couple of Facebook videos and I like taking pictures too – Dad says I'm good and it annoys him when my photos get more likes on Facebook than his!"

Clare

"I get involved with Richard's cooking demos on a Wednesday.

I didn't start out to be the mystery voice in the proceedings, it's just the way it's turned out. There's the odd comment on social media, 'Why don't we see Clare?' and so on, but I don't work on the farm, I'm just happy to support it as it's part of my family now. I also like being able to nip into the farm shop and not be recognised!"

Anita

"I am the merchandising manager and it's really taken off since the boys have been on television. When I'm doing phone bookings it's funny when people know I'm Dave's wife, but I don't like having my photo taken, so there aren't many pictures of me, and I like being a kind of mystery!"

Tom

"It was a bit surreal when Dad and David started appearing on TV, even though I'd seen him on lambing reports in the past. We always knew he was a great storyteller and I'm just really proud that he can show off the side that we have always seen. He's very confident and relaxed and lets his true self shine through and I think that's why people respond to him."

Clare

"I love the fact that Cannon Hall Farm is serving South Yorkshire so effectively; the landscape is gorgeous here. It's still a working farm and Roger will still go out every day and check his animals are alright, so essentially his day-to-day role hasn't changed to the one he had 50 years ago. And that's what makes the place so special – this mix of the traditional and the very modern."

Katie

"Friends at school used to say, 'What's it like having thousands of strangers on the farm all the time – it must be so weird?' But when I went to live in a 'normal' house with my husband, it wasn't my normal – that's to be surrounded by tourists and shoppers. It's been the most incredible place to grow up in, and my daughter Nelly is loving it, too. I can't wait for her to ride Jon Bon Pony!"

25

What's next for Cannon Hall Farm?

Looking ahead to the next few years, the farm will continue to expand with new animals and new facilities for them to live in, and new members of staff coming to join the happy gang – no doubt on a part-time basis that will somehow turn into several years in the blink of an eye.

Followers of Robert's Facebook Live broadcasts will have seen new buildings being created on the farm and there are big plans afoot for the Small Mammal Building. "We're planning a really good variety of both native and exotic small animals to wow our visitors," says Robert.

Along with common varieties of mice, rats, river fish and guinea pigs, the farmers are looking forward to welcoming Gambian Pouch Rats. As their name suggests, these African rodents have large cheek pouches like a hamster that they stuff with food. Although they are called rats, they are much bigger rodents and have been used to detect both land mines and tuberculosis because of their extraordinary sense of smell. "We've found that other wildlife collections are happy to share some of their animals," says Robert. "They can't sell them to the pet market because that's not deemed acceptable. A bad life is worse than no life at all and something not kept to the right enrichment and health standards would be better off dead than living a life of purgatory."

"I'm excited about getting Sugar Gliders which are marsupials that can jump from tree to tree," says David. These small noctur-nal squirrel-like possums glide through the air with the help of membranes between their front and back legs that act as a parachute as they leap through the trees. They love sugary foods, hence their name, and have huge eyes and swivelling ears that are adapted for night vision.

Red squirrels are also earmarked for a luxury suite in the new small mammal zone. "There are still a few enclaves in the UK where red squirrels can be found in the wild," says Robert. "But we are getting ours from the Welsh Mountain Zoo. They are from a stud book, so it's known exactly who the father is, so there's no possibility of interbreeding which weakens the breed. What they are trying to do is to keep a pool of red squirrels just in case they do die out in the wild."

Angora Rabbits were a feature of the farm when it first opened to the public and more bunnies are in the new plans, only different varieties, such as the unbelievably cute Netherland Dwarf rabbit and Giant rabbits. "The Small Mammal enclosure is going to be one of the nicest areas of the farm and we're really excited about it," says David. "Depending on restrictions moving forward, not all animals will be able to be handled, but we're hoping to reintroduce some of the elements of a petting zoo, so visitors can get hands-on."

Outside in the farmyard, visitors will be able to spot owl boxes dotted around for the farm's Tawny Owl population. "We're also planning a nice rich meadow up in the wood to encourage inver-tebrates, shrews and other birds to build up the food chain," says David. Badgers, bats, deer and owls can already be found in Deffer Wood and the Nicholsons are keen to get more visitors out walking and enjoying native breeds.

To this end, they are also planning a big extension to the White Bull Restaurant, to encourage more local dog walkers to come along with their four-legged friends. What is currently the Lucky Pup dog-friendly cafe may turn into the merchandise zone.

"Everything is happening so fast on the farm, we're continually making plans," says Robert. "COVID-19 threw a huge spanner in the works for 2020 and a lot of the plans that we had been making had to be put on a back burner, but we've been able to dedicate more time to filming and raising the profile of the business, so hopefully that puts us in a good place when life returns to normal again. The fact that we are continuing to work with really nice people around us and trying things that we have never done before is wonderful. Getting to fifty-something and trying something new is always good, isn't it?"

Acknowledgements

The Nicholson Family and Nicole Carmichael would really like to thank the following people for all of their help writing this book.

First and foremost, to Amanda Stocks, for helping make the idea of the book a reality, and to editor Jo Sollis and all the team at Mirror Books for helping share the Nicholsons' story.

Thanks to Paul Stead and everyone at Daisybeck Studios for the opportunity to show the Nicholson family farm to a wider audience and to Ben Frow, Sebastian Cardwell and Dan Louw at Channel 5. Thanks also to Peter Wright, Julian Norton, Jules Hudson, Helen Skelton, Adam Henson, JB Gill, Kelvin Fletcher, Matt Smith, David Melleney and Shona Searson.

Thanks to Steve Wilson for all of his help and research into the Nicholson dynasty, Beryl Robertshsaw for her memories of life at Bank End Farm and Rosemary Brain for her hilarious tea room tales and stories of family holidays, carnival fun and any time there was an opportunity to party.

Thank you to Nicola Hyde and all the amazing staff and family members who gave up their time to be interviewed for this book. Your dedication to the continuing success of Cannon Hall Farm is extraordinary. Also a very heartfelt thank you to the thousands of staff and millions of customers who have shared the Nicholsons' journey over the years. Some are no longer with us, but all made a valuable contribution to making the farm the success it is today, the family will be forever grateful.

Roger would especially like to thank John Linford for his help in looking back to childhood days and for reminiscing how life was in those early years, Ron Carbutt for his advice and assistance when opening the farm to the public and to Will, David, Alan and Cynthia Roe for their help throughout his life.

Thank you to Rosemary and Nigel Brain for their help with haymaking, harvesting, turkey plucking and their crucial help when the children were young. To Cynthia's parents Olive and Edward for their support through thick and thin and to Cynthia's brother Ted. Roger says: "Finally, thank you to my dear wife Cynthia who has been such a strength in our marriage and never let me down. Much love to her."

Nicole would personally like to thank Roger, Cynthia, Richard, Robert and David for making the book an absolute pleasure to write. You are a fantastic family and as genuine, funny, kind and real as you appear on screen. Your children and other halves are all very special too. Thank you so much.